A REFLEXIVE READING OF URBAN SPACE

Providing a critique of the concepts attached to the representation of urban space, this ground-breaking book formulates a new theory of space, which understands the dynamic interrelations between physical and social spaces while tracing the wider urban context. It offers a new tool to approach the reading of these interrelations through reflexive reading strategies that identify singular reading fragments of the different spaces through multiple reader-time-space relations. The strategies proposed in the volume seek to develop an integrative reading of urban space through recognition of the singular (influenced by discourse, institution, etc.); and temporal (influenced by reading perspective in space and time), thereby providing a relational perspective that goes beyond the paradox of place in between social and physical space, identifying each in terms of relationships oscillating between the conceptual, the physical and social content, and the context. In conclusion, the book suggests that space/place can be read through sequential fragments of people, place, context, mind, and author/reader. Operating at different scales between conceptual space and reality, the sequential reading helps the recognition of multiplicity and the dynamics of place as a transformational process without hierarchy or classification.

Mona A. Abdelwahab is an Assistant Professor in Architecture, Department of Architecture and Environmental Design at the Academy for Science, Technology and Maritime Transport, Egypt. She received her PhD in Architecture from Newcastle University, UK, where she is a visiting research fellow. She followed her post-doc studies at the Department of Spatial Planning, University of Groningen, NL, where she co-founded 'YA-AESOP- Booklet Series: Conversations In-Planning'. She is also cofounder and managing editor of *'Arcplan': Arabic cities planning* e-journal.

New Directions in Planning Theory
Series Editors: Gert de Roo, Jean Hillier,
Joris Van Wezemael

The series New Directions in Planning Theory develops and disseminates theories and conceptual understandings of spatial and physical planning which address such challenges as uncertainty, diversity and incommensurability. Planning theories range across a wide spectrum, from questions of explanation and understanding, to normative or predictive questions of how planners should act and what future places should look like. These theories include procedural theories of planning. While these have traditionally been dominated by ideas about rationality, in addition to this, the series opens up to other perspectives and also welcomes theoretical contributions on substantive aspects of planning. Other theories to be included in the series may be concerned with questions of epistemology or ontology; with issues of knowledge, power, politics, subjectivation; with social and/or environmental justice; with issues of morals and ethics. Planning theories have been, and continue to be, influenced by other intellectual fields, which often imbue planning theories with awareness of and sensitivity to the multiple dimensions of planning practices. The series editors particularly encourage inter- and trans-disciplinary ideas and conceptualisations.

Published:

The Paradoxes of Planning
A Psycho-Analytical Perspective
Sara Westin

Making Use of Deleuze in Planning
Methodology for Speculative and Immanent Assessment
Gareth Abrahams

Decentralization in Environmental Governance
A Post-Contingency Approach
Christian Zuidema

A Reflexive Reading of Urban Space
Mona A. Abdelwahab

A REFLEXIVE READING OF URBAN SPACE

MONA A. ABDELWAHAB

Routledge
Taylor & Francis Group

LONDON AND NEW YORK

First published 2018
by Routledge

2 Park Square, Milton Park, Abingdon, Oxfordshire OX14 4RN
52 Vanderbilt Avenue, New York, NY 10017

Routledge is an imprint of the Taylor & Francis Group, an informa business

First issued in paperback 2020

British Library Cataloguing-in-Publication Data
A catalogue record for this book is available from the British Library

Library of Congress Cataloging-in-Publication Data
A catalog record for this book has been requested

ISBN: 978-1-4094-5228-7 (hbk)
ISBN: 978-0-367-59261-5 (pbk)

Typeset in Warnock Pro
by Apex CoVantage, LLC

To Jean Hillier and Peter Kellett.

CONTENTS

ILLUSTRATIONS

FIGURES

TABLES

ACKNOWLEDGEMENTS

The journey to write this book has been across three universities – Newcastle University, UK, Cairo University, Egypt, and Groningen University, NL. It would not have been possible to accomplish without the valuable guidance and support I have received in preparation and completion of this study from many people who influenced and contributed to both my life and work.

I owe my deepest gratitude to Prof. Jean Hillier, who has been my mentor and inspiration through this journey. She helped me to realise and develop my abilities as both researcher and academic and continued to encourage and support me through every difficulty. I am also truly indebted to Dr. Peter Kellett, who believed in and supported my work long before I did. He has always been there for me at every turn about or obstacle I had to walk through, which is a key factor in this achievement. I am heartily thankful, and very proud to have been their student, a relationship that extended beyond work to touch my heart and stay there forever.

I would like to take this opportunity to convey my sincere thanks to Prof. Patsy Healey, not only for the advice that helped shape the vignette of 'The Cultural Park for Children' in this book, but also the continuous and inspiring discussions that motivated me. I owe the earnest thankfulness to Prof. Gert de Roo for the opportunity to do my post-doc, which contributed to the preparation of this manuscript. I extend my gratitude and sincere thanks to Prof. Ali Gabr and Prof. Basil Kamel, who have made available their support in many ways particularly at the start of this work.

Many thanks for the Aga Khan Trust for culture, especially media and documentation office Lobna Montasser, for the permission and provision of the photos of 'The Cultural Park for Children' used in Chapter 4. I would like to acknowledge that the first part of Chapter 3, 'Who is Cairo?' was previously published as 'Cairo: A Deconstruction Reading of Space' in Kellett P. and Hernández-Garcia (eds.), 'Researching the Contemporary City' (2013). I extend my thanks to director of Javeriana Publishing House, Nicholas Morales Thomas, for providing the permission to re-publish.

It is a great pleasure to thank everyone who helped me. Many thanks to the publishers, especially editorial assistant Valerie Rose, Priscilla Corbel and Faye Leerink for their assistance, guidance and repeated extension of deadlines over several years. And special thanks to editorial assistant Rebecca McPhee for her caring and conscientious revision of every detail towards the delivery of the manuscript. My thanks also extend to Bernadette Williams for the exceptional care in proofreading an earlier draft of this manuscript.

Lastly, I am much obliged to my parents for their continued support and encouragement.

A THRESHOLD

'A reflexive reading of urban space' is conceived through four instances: a question, an observation, a quotation, and an affiliation, each of which is a reflection on urban space. These instances help the recognition of the complexity and dynamics of urban space, and highlight the gap in theory to approach these complexities that implicate the design and planning of new urban spaces. This 'reading' thus questions the different approaches to reading urban space in architecture, social studies, and philosophy. It is a theoretically based study of urban space in general, which is developed through a thinking space in Cairo, Egypt. However, the study of a specific urban space is not the focus of this book as such, but the way these spaces are read, understood, and accordingly planned and represented. This book explores the relation between the reading and interpretation of urban space on the one hand and its construction on the other. However, it is not concerned with the planning process as such. This book thus formulates a 'reading space' beyond the traditionally prevalent readings and representations of urban space, it seeks continuous exploration of urban space through a post-structuralist journey that involves visual-thinking diagrams, multiple theories, ~~crossing out~~, as well as questions the role of philosophy. Finally, this reading unfolds the theme of reflexivity, which supports the continuous reflection between the different approaches and simultaneously continues to avoid and question the dominance of a theoretical and/or interpretive perspective. However, the notion of 'reflexivity' is elaborated further through the following sections.

FOUR INSTANCES

I was inspired to write this book by four instances that do not follow any particular order. These instances involve a question: what if Derrida were an Egyptian – taken after 'Mosaic Fragments: If Derrida Were an Egyptian' (Bennington, 1992). This instance arises from a consideration of the visit of the French philosopher Jacques Derrida to Cairo in 2000. This visit developed a considerable deviation from the consensus reading of Cairo in the literature; Derrida's reflections on the city, its image, history, and social realities, and what he described as the misrepresentation of the city holds the potential for an alternative reading. At the same time, a quotation by the architect and theoretician Bernard Tschumi (2001) is highlighted, emphasising the hybrid reality of architecture. This quotation helps in defining architecture in between idea and reality, architectural concept and social realities, and intrinsically extends to urban space. Simultaneously, this is complemented by an exploration and observation of Cairo space, which reflects the multiplicities, chaos, and conflict between Cairo and Cairenes. The latter is the English pronunciation of the Arabic word (Caheryeen – قاهرين) and is used by many ethnographers and writers to refer to the people of Cairo. This raises the question about my role as an architect in perception of this space. This question helped to develop the final instance; my personal affinity towards theory and philosophy, which was deepened by the evident lack of literature on theory concerning the region. I shall hence trace and explore these instances in this section.

A question: what if Derrida were an Egyptian?

> Hieroglyphs and pyramids, Thot and Isis, colossi and the Sphinx: Egypt repeatedly returns to haunt Derrida's writings. From the two (or three) great Plato readings to the great Hegel readings, via discussions of Freud and Warburton, Egyptian motifs regularly appear at important moments in the texts.
> What is the place of Egypt in deconstruction?
>
> (Bennington, 1992: 97)

Jacques Derrida, the French philosopher, was born and brought up in Algeria (Bennington and Derrida, 1993). His name is forever associated with his deconstruction project (Collins et al., 2005; Hill, 2007). Accordingly, his work is controversial; it is considered both the 'most significant in contemporary thinking' and a 'corruption of all intellectual values' (Collins et al., 2005: 1). Derrida visited Cairo in February 2000 where he gave a series of lectures at the Egyptian Supreme Council for Culture (Al-Ahram-Weekly, 2000a). These lectures and his reading of Cairo were predictably controversial, and accordingly developed a strong opposing debate. This section presents a review of Derrida's lectures about Cairo, his ideas and comments, which questions the traditional reading and representation of Cairo urban space.

> Unhappy he who claims to be his own contemporary.
> Derrida does not: I would imagine him, rather, with Plato and a few others, at Heliopolis,[1] in Egypt.
>
> (Bennington and Derrida, 1993: 8)

On his visit to Cairo, Derrida gave a series of lectures, where he engaged with the Egyptian audience through a representation of the perspective and conceptions of his work as well as through his reading of Cairo, and Egypt in general. It was rather difficult to find the script of Derrida's lectures, which was archived neither in western publications nor in Egypt. Accordingly, I resorted to newspaper articles as well as an article published in an Arabic journal of philosophy as a secondary source of Derrida's visit to Egypt. Furthermore, the lectures were given in French; although many Egyptians are well acquainted with the language, it is the English language that is more widely understood. Despite this, this event attracted many Egyptian academics and intellectuals. However, while specialists were invited and 'some 50 observers selected', the event was structured so as to put off the 'non-specialists' (Al-Ahram-Weekly, 2000b) and were held away from the university and the students. On the last day of his visit, Derrida explicitly expressed his regret at not having met any students (Al-Ahram-Weekly, 2000b). The context of these lectures and the audience involved made the discussion and representation a reflection of the conventional and traditional stream of thought within the political and intellectual parties.

These lectures therefore reflected on philosophy, deconstruction, and the humanities in general, and more specifically with reference to Egyptian identity, especially through the lecture 'Egyptian References: Origin, Orientalism and Theory of Deconstruction'. With reference to the Egyptian context, Derrida reflected on the concept of hospitality, 'the difference between the conditional and unconditional hospitality', and the multiplicity of the Egyptian/Cairene cultural identity (Al-Ahram-Weekly, 2000b). Accordingly, he discussed 'the meaning of place', and the psychological complexity' of the people in relation to the Egyptian/Cairene context, and its geographical and historical setting. In response, the audience questioned Derrida about the legitimacy and relevance of deconstruction to the Egyptian socio-political context; and more specifically its relevance to the historic context (Al-Ahram-Weekly, 2000b). The majority considered deconstruction 'peripheral' to the Egyptian context and the Arab world in general (Rakha, 2001). Their questions 'retained an emphasis on the dominance of Western thought' on a 'post-colonial' culture and stressed their attempts to integrate with the Arab culture, 'making it all part of us' (Al-Ahram-Weekly, 2000b). This highlights the dominance of the image of the city as part of the Arab world, a form of regionalism. Al-Messirri (2005) expresses several reservations concerning Derrida's perspective of the multiple identities of Cairo if these multiplicities meant that Cairo has no identity, or has lost, or is losing her identity.

In response Derrida moderated his philosophical perspective saying, 'You know, I've only been here three days!' as he explained that the audience was not obliged to 'engage with deconstruction' if it was not part of their beliefs. Interestingly, the relationship between Egypt and Derrida is neither new nor initiated by his visit to Egypt in 2000. Derrida's interest goes back to the dialogues between the Egyptian writer in ancient Egypt and Plato. He further explained that he was not a dreamer of utopia. He did not support globalisation – as opposed to nationalism – either, but displaced 'globalisation' with 'mondialisation', a sub-term of internationalism, which implies a crossing and blurring of the boundaries between nations rather than their unification through globalisation.

An observation: a vignette from Egypt

Fatimid Cairo, the Egyptian capital, was founded in AD 969 by Gawhar Al-Sekly. Later, Al-Muez Ledin Allah Al-Fatimy, the Fatimid leader, called the City Al-Qahira, the city victorious. Today, the city of Cairo has grown into a high density, overcrowded metropolitan city with a population of over 12 million inhabitants and a density of over 40,000 persons per square kilometre (Ethelston, 2016). Although the metropolitan area of Cairo has increased significantly, the urban public spaces, mostly referred to as green areas, have notably diminished (El-Messiri, 2004). Built on former agriculture and desert land, the city constitutes a densely laid urban fabric with few spaces in between (Rabbat, 2004). Simultaneously, the last two decades have witnessed a growing concern, on both the national and international level, to relieve Cairo of pollution and environmental and social stresses. This was complemented by an increasing academic interest in the study of Cairo urban space, evidenced by the launch of project initiatives to approach urban space. I hence explore the development of urban space in Cairo through the observation of the Cairenes' relationship to public space, the Cairenes' perspective towards formal public spaces as well as their adaption to these and other more informal spaces (Abdelwahab, 2010).

Cairenes: the people of Cairo

> Cairo . . . is overcrowded with people of vastly differing backgrounds, heterogeneous cultural values and rapidly changing class structures.
>
> (El-Messiri, 2004: 221)

The people of Cairo, the Cairenes, provide another milestone in the story of Cairo's public space: 'Crowding squeezes Cairenes out of their homes. But where to go? There are precious few green spaces' (Rodenbeck, 2005: 17). However, the few existing public spaces play a vital role in the public life of Cairenes. Scholars consider that the introduction of the concept of public space/green areas in the Cairene society is relatively new (El-Messiri, 2004). Controversially, Fatimid Cairo was developed around the garden of 'Kafur' (Rabbat, 2004). Today, part of Cairo's public space, with special reference to the historic city, has been developed as a tourist attraction, and many spaces are fenced keeping Cairenes outside (Singerman and Amar, 2006a; Abaza, 2006). Fenced spaces include both some touristic spaces as well as small green islands. This contradicts the scholars' conception that public spaces are relatively new to Cairene society. It is apparent that Cairenes need and appreciate the life of public spaces. However very few spaces are accessible to the public owing to both the scarcity of urban spaces, and the exclusion of Cairenes from many places through regulations or socio-economic segregation, which is both evident and rooted in the history of the city. Accordingly, Cairenes are struggling to reclaim their own public space.

All Cairenes objectively consider distinctions between classes of public space in their discourse. The wealthiest say they do not mix with the

poorest of society, because they fear crime, insults, filth. . . . The popular
classes say they fear feeling out of place, standing out.

(Battesti, 2006: 503)

The diversity of the Cairene/Egyptian social culture helps to raise the question
'whose Cairo?' which is the title of an article by Al-Sayyad (2006). Reflecting
on the Cairene's relationship to their urban public space, three social categories of public spaces can be identified: private, public, and informal. The
social distinction or disjunction between the private and the public is evident within the socio-culture structure of Cairo (Battesti, 2006; El-Messiri,
2004; Singerman and Amar, 2006a). On the one hand, the rich are present
in exclusive clubs, gated communities, and private gardens – isolated from
the public. On the other hand, the users of public gardens/space represent
the poor social classes (Battesti, 2006; El-Messiri, 2004). Another interesting phenomenon is the Cairene approach to informal settings. They look at
other informal settings, such as sidewalks and squares, rooftops, bridges,
and so on and turn them into places for commerce, entertainment, play-
grounds, open-air mosques, cafes and restaurants, special ceremonies (Battesti, 2006; Rodenbeck, 2005; Salama, 2004; Singerman and Amar, 2006a).

A quotation: architecture-urban space

Indeed, architecture finds itself in a unique situation: it is the only
discipline that by definition combines concept and experience, image
and use, image and structure . . . architects are the only ones who are
the prisoners of that hybrid art, where the image hardly ever exists
without a combined activity.

(Tschumi, 2001: 257)

It is natural, since I am an architect, to involve an instance that reflects on
architectural space: a reflection through Cairo space. This instance is inspired
by the work and writing of the post-structuralist architect and theoretician
Bernard Tschumi. It is worth noting here that Tschumi and Derrida worked
in collaboration, together with Peter Eisenman, in Parc de la Villette, Paris
on the project that momentarily brought together deconstruction (through
Derrida) and architecture theory (through Tschumi and Eisenman), and
introduced a long open conversation between the two approaches.

The definition of architecture . . . must expand to an urban dimension.
The complex social, economic, and political mechanisms that govern the
expansion and contraction of the contemporary city.

(Tschumi, 2001: 22)

Post-structuralist architects, and Tschumi in particular, are pre-occupied
with the disjunction between the abstract idea and the social realities of
place: concept/experience, theory/practice, space/activity, image/use, and
so on (Kamel and Abdelwahab, 2006). Tschumi (2001) considers architecture to be a prisoner in between these disjunctions. He perceives architecture space as being in between the abstract concept and the experiences of

▶ **FIGURE T.1**
Architecture
space in between
concepts and
experience, in the
urban context

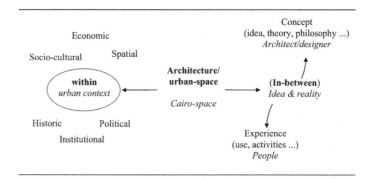

people in the use and activity within the space (Figure T.1). Simultaneously, he considers the fundamental expansion of architectural space within the urban context, social, economic, political, and so on. Accordingly, as shown in Figure T.1, architecture/urban space operates on two levels, in between abstract idea and reality, and in relation to its wider context. I thus project this reading to Cairo space.

An affiliation: theory/conceptual thinking

> In fact, no comprehensive social science examination of contemporary Cairo has been published since Janet Abu-Lughod's *Cairo: 1001 years of the city victorious . . .* drafted in the 1960s at the high point of the turbulence and optimism of Gamal Abd Al-Nasser's revolution.
>
> (Singerman and Amar, 2006a: 21)

My personal affiliation helped to formulate the book theme through a theoretical rather than an empirical study. This affiliation was reinforced by the evident lack of a body of literature in architecture theory in Egypt as well as the Arabic region. As discussed, there has been a concurrent interest in the development and study of Cairo urban space, and simultaneously, the dilemma of urban space, identified as spatial, historic, social, cultural, political, economic, and institutional, is explored. However, these studies mainly considered particular dimension(s) rather than the relations between them; so reflecting only one side of Cairo, whereas Cairo space shows as divergent, dynamic, and continuously changing. These studies of Cairo urban space also show the dominance of the historic dimension. Elsheshtawy (2004: 2) demonstrates the dominance of 'a singular and short-sighted theme – namely socio/religious reading of urban spatial patterns', which linked the city to reflections and pre-conceptions detached from the surrounding reality; and accordingly, influenced the emerging relations and patterns of power in Cairene urban space. This dominance therefore resulted in a recurrence of the polarisation conflict between Islamic concepts, tradition, and nationalisation on the one hand and western concepts, modernisation, and globalisation on the other. Elsheshtawy (2004) also acknowledges the development of the city through the modernisation process, which needs to be explored further. However, these studies of Cairo, as with other 'traditional Islamic cultures' lack a comprehensive theory of urban space (Hakim, 1999).

BEYOND PREVALENT READINGS OF URBAN SPACE

These four instances hence reflect on two main issues. The first considers the paradox of Cairo urban space in between Cairenes and Cairo: the social and the physical spaces. This paradox has arguably developed through the inherited conceptions and/or misconceptions of her identity and representation, which has led to a further alienation and misreading of her true self. Derrida's reading of Cairo space recognised this mess, and challenged these conceptions. He further reflected on the unseen, ignored multiplicities, complexities, and potentials through a monolithic historic image misrepresentation. And the second reviews the inherited paradox of architecture between idea and reality, which simultaneously extends to the wider urban context. It also considers the lack of theoretical body of work in the region to help in approaching this dilemma, a missing area in theory to address the urban space and context.

Taken together, these four instances help to highlight architecture and urban theory in correlation with philosophy, particularly Derrida's deconstruction project, on the one hand and with social studies that consider the empirical study of urban space on the other. The question raised is: 'How to' approach, read, and understand the paradox of urban space in between Cairene and Cairo space? This question intrinsically seeks a theoretical contribution. It is not concerned with an investigation of the identity of Cairo space, but rather to investigate 'how to' investigate this identity. It hence formulates a double-sided reflection on the paradox of Cairo space through the study of the paradox of architecture and urban space (see Figure T.2).

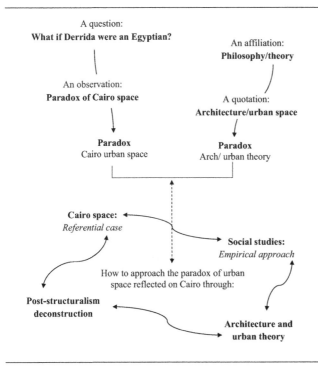

◄ **FIGURE T.2**
The 'paradox' of urban space in reflection to Cairo space

Consequently, this book seeks to formulate a new theory of urban space, which understands the dynamic interrelations between the social and physical spaces while tracing the wider urban context: political, economic, and so on. It hence provides a critique of the concepts attached to the representation of urban space, reflecting on theories and philosophies as well as social studies in order to approach the study of Cairo urban space.

The main aim is to develop a multidisciplinary framework of urban space, with special reference to Cairo space that operates on two levels: urban space/place and context. Although, the use of a 'framework' implies the development of a consistent analysis to provide a well-defined classification and clear image of urban space, in fact this is an attempt to read the not-well-defined, the inconsistencies, and misreadings of a rich, complex literature of urban space and its everyday realities. Accordingly, the expected framework would not provide a consistent reading or analysis strategy but would be expected to lead to strategies for approaching this paradox.

A REFLEXIVE READING OF URBAN SPACE

This book constitutes two main parts: a preliminary and a reflexive reading of urban space, and three interludes to set up these reading spaces: 'A threshold', 'A turning point', and 'To be continued . . .' (see Figure T.3). 'A threshold' instigates the reading and writing of 'urban space' in this book in

▶ **FIGURE T.3**
A reflexive reading
of urban space
outline

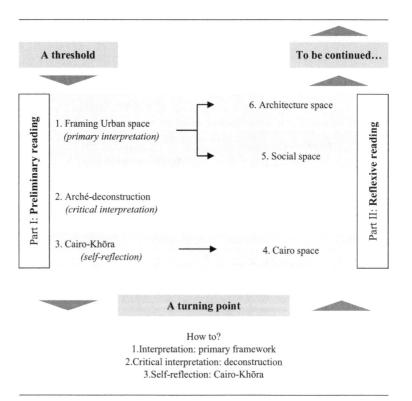

general. It simultaneously highlights the implications of working within a post-structuralist theory, particularly the setting for continuous exploration of urban space, 'To be continued'.

The 'preliminary' reading, in Part I, thus reviews three spaces. The first chapter explores multiple disciplinary approaches to urban space through philosophy, architecture theory, and social studies. This chapter explores similarities and inconsistencies in the literature and introduces a primary framework to approach urban space. The second explores the deconstruction project as reading strategies, and its resonance in architecture theory. It is worth noting that this book is interested in the study of deconstruction in relation to architecture and urban space as a way of thinking, particularly the concepts of space/place rather than as a 'performative art' – a style. Finally, the third chapter explores two reading vignettes of urban space, Cairo and Khōra, in reflection to deconstruction.

'A turning point' is the intermediate interlude that revisits the reading(s) of 'urban space' in response to the preliminary reading, in reflection to the multidisciplinary nature of urban space and the difficulty in approaching urban space through a structured 'framework'. This 'turning point' thus adopts the notion of 'reflexivity', which involves the capacity to re-approach the multiplicities and dynamics of urban space through a continuous interplay between the various data introduced: both theoretical and empirical, without any of them becoming dominant.

Part II thus turns back to re-approach the three reading spaces drawing on 'reflexivity'. This book adopts Alvesson and Skoldberg's (2000) quadrihermeneutics example of reflexive interpretation, four levels of interplaying interpretation: data construction, primary interpretation, critical interpretation, and self-reflection. The preliminary reading enables the data construction of social and architecture space to be re-approached in Chapters 5 and 6 respectively, as well as helps the choice and development of the Cairo vignette to be re-approached in Chapter 4. Primary interpretation represents the basic level of interpretation, through the framework of urban space introduced in Chapter 1. Critical interpretation principally questions and challenges the prime interpretations, through deconstruction-reading strategies introduced in Chapter 2. Finally, the vignettes of Cairo and Khōra are designated as self-reflection themes, which question the interpretation process through reflection on interpretation authority and selectivity.

At the same time, the reflexive approach is elaborated and further developed in the course of reading the three spaces; the main objective is to develop the primary 'framework' presented in Chapter 1 into reflexive reading strategies of space. These reflexive spaces continue to negate and destabilise the established primary framework of space, demonstrating the conflicts and inconsistencies as well similarities. Finally, these spaces reflect on each other on the margin of the reading space.

The third interlude reflects on the book's journey to explore the reading of urban space. It thus emphasises the singularity of every reading space that echoes the perspective of various authors/readers, and the implication of reading space as dynamic relations rather than the social, the physical, and so on and attempting to relate them. Accordingly, it proposes the reading of urban space through various fragments presenting the sequences and constituents of urban space. However, this new reading of urban space must remain without a conclusion, to be continued . . .

A POST-STRUCTURALIST JOURNEY

> Instead of an integrated theoretical frame of reference which guides an analysis towards unequivocal, logical results and interpretations, the idea is to strive for multiplicity, variation, the demonstration of inconsistencies and fragmentations, and the possibility of multiple interpretations.
>
> (Alvesson and Skoldberg, 2000: 152)

The involvement of the deconstruction project implies the adoption of post-structuralism. However, it should be noted that the journey of this book is also embedded within a structuralist notion. My personal experience is deeply rooted in a structuralist frame of reference. This journey helped me to recognise these structuralist tendencies, and to understand and develop a post-structuralist frame of reference. Significantly, the book subject – the reading of urban space – also developed from a structuralist reading to a post-structuralist one. Simultaneously, the selection of theories in this book includes both approaches; for example, architectural space oscillates between Markus's structuralist and Tschumi's post-structuralist reflections. In general, post-structuralism negates structuralist approaches and inherited western metaphysical ideas (Alvesson and Skoldberg, 2000; Groat and Wang, 2002). A traditional/structural approach attempts to 'isolate elements, specify relations and formulate synthesis' to reach a universal meaning/truth (Rosenau, 1992: 8). A post-structuralist approach attempts to do the opposite, 'indeterminacy rather than determinism, diversity rather than unity, difference rather than synthesis, complexity rather than simplification' (Rosenau, 1992: 8), to demonstrate that there is no best interpretation and no singular meaning/truth (Alvesson and Skoldberg, 2000).

> However, there is no end to the task of deconstruction.
>
> (McQuillan, 2001: 15)

Controversially, Alvesson and Skoldberg (2000) present a critique of post-structuralism. A post-structuralist approach works from within another approach to problematise and expose the weak points within, to deconstruct. Accordingly, post-structuralism, and simultaneously deconstruction, lacks the ability to construct/reconstruct another discourse. It risks an over-simplification of the other perspective through the continuous attack. It also risks a linguistic and textual reduction of the subject, which then detaches it from socio-cultural realities. Finally, Alvesson and Skoldberg (2000) consider the problematic possibility of the overemphasis on these linguistic and textual strategies, which strive for multiplicity and variations in both interpretation and representation in such a way as to jeopardise the content. They called this 'the Sokal affair'.[2]

The difficulty of working within a post-structuralist context could be reflected in this book in three ways. The first considers the affinity of post-structuralism to continuous exploration. There is a potential risk to be trapped in the data collection phase and turn the book into merely a substantial literature review of the subject. The second considers the subject, the urban space, as intrinsically embedding a social phenomenon, which cannot and should not be confined to textual and a linguistic analysis. Finally, the process of questioning, destabilising, and/or deconstructing a discourse should be taken into consideration.

> It is not method, but ontology and epistemology which are the determinants of good social science.
>
> (Alvesson and Skoldberg, 2000: 4)

A common approach to the classification of systems of inquiries defines these as independent spheres of positivist/post-positivist, naturalistic, and emancipatory, depending on which ontological and epistemological assumptions are referenced, as well as the measures of quality used. The three paradigmatic clusters were introduced and adopted by Groat and Wang (2002). However, this classification includes basically contradicting ontological and epistemological assumptions that vary from an objective perspective of a single truth of one reality, to an identifiable subjective perspective of multiple truths and realities situated in its local context. This contradiction makes their clustering rather problematic.

Significantly, Alvesson and Skoldberg (2000) introduced a complementary classification in *Reflexive Methodology*, which defines these systems of inquiry as data-driven, insight-driven, and emancipation-driven strategies. A data-driven study considers the careful interpretation and construction of empirical material. An insight-driven study emphasises a hermeneutic process that implies an in-depth understanding of the text – the empirical material. A critical emancipation-driven study on the other hand, goes beyond the empirical data construction, in favour of 'reflective critical observations' of the subject. This latter approach is deconstruction. This classification helps the interplay between the three strategies, according to the approach and nature of the discourse. A desirable situation would be the 'combination of extensive empirical work and advanced critical interpretation' (Alvesson and Skoldberg, 2000: 257–258). Following this analysis, this book works mainly through an emancipation-driven strategy which provides a flexible and dynamic frame of reference and helps the reflexive approach to the empirical material. It should be noted that empirical material involves theoretical considerations as well as other fieldwork data.

Consequently, we shall explore the role of post-structuralist philosophy in the context of this book. It demonstrates the book's continuous exploration through visual-thinking diagrams, representation of different and sometimes conflicting voices/theories: to avoid and question the dominance of a particular theoretical perspective – this also supports reflexivity as will be elaborated later – and lastly, the ~~crossing-out~~ process which illustrates the development of ideas and themes related to theories of space rather than delete or reject.

The role of philosophy

Groat and Wang (2002) consider philosophy as a background for the legitimacy and coherence of theory, and necessary in order to explain the nature or socio-cultural setting. They hence introduced a linear chart initiated by philosophy towards theory and strategy. On the other hand, Alvesson and Skoldberg consider the role of philosophy – namely post-structuralist and critical theory – as a meta-theory. They demonstrate that meta-theory is non-static and relational and not directly involved with empirical material interpretation but rather with the 'guides and frames' of theoretical interpretations. Meta-theory has two roles: 'to problematize the legitimacy of dominant interpretive pattern', and to provide potentialities for new interpretations. They are sceptical about the use of a particular philosophy as both a theory and a meta-theory, because it raises the fear of 'theory-reductionism' and biased interpretation as well as the need to use a different insight in order to reflection on one another (Alvesson and Skoldberg, 2000: 253–254, 291).

However, in this book post-structuralism, and particularly deconstruction, plays both roles. As discussed, this book is primarily developed within a post-structuralist context. Accordingly, deconstruction is involved in the development of the theoretical approach of urban space presented in the preliminary reading. This is manifested through the concept of Khōra, the deconstruction reading of space, and Tschumi's approach to architecture/urban space, among others. In this sense, the role of post-structuralism/deconstruction in relation to the book subject could be seen as both immanent and transcendental. Post-structuralism is immanent, in the sense that it works from within the constructed subject to formulate the interpretations. And deconstruction is transcendental, in the sense that it provides a reflection on the framework used in the interpretation without being part of the book's empirical material and interpretation. Accordingly, to avoid 'theory-reductionism', and manage this paradoxical role of post-structuralism, the book is structured and developed in consecutive reflexive instances. These instances help to isolate the two roles of post-structuralism, as part of the empirical material and as a meta-theory.

Multitransdisciplinary reference(s)

This book approaches urban space between idea – the conceptual space in philosophy and theory – and reality – the socio-spatial embedded in the contextual in architecture, urban design, and social studies. These three disciplines are not isolated but interact through their approaches to urban space, and accordingly produce many similarities and differences, agreements, and conflicts. Interestingly, post-structuralism is pre-occupied with the disjunction between the idealist philosophy of space/place and the realities of social space (Peter Eisenman interview in Solzhenitsyn, 1998). However, the epistemological and ontological differences between these two approaches are worth noting. Post-structuralism explores the phenomena of urban space through theory and conceptions, while the social studies aim to understand the phenomena of urban space through the construction of social space. Traditionally, social studies, particularly environmental psychology, are associated with positivism and follow an empiricist approach (Broadbent, 1988; Groat and Després, 1991; Lang, 1987). Simultaneously, this traditional approach helped to separate social studies from architecture and urban theories and practice, which developed in association with postmodernism/post-structuralism (Groat and Després, 1991). However, social studies drifted into naturalistic approaches, which associated itself with explanatory theory and accordingly developed a closer relation to architecture and urban theory (Bell, 2001; Lang, 1987). Finally, social studies developed within a postmodern context and considered the experience of urban space through 'epistemological uncertainty and ontological plurality' (Best, 2003: 270); it shifted towards post-humanism and post-foundationalism. In addition, models/theories of place were developed within social studies through various pathways through a positivist empiricist approach (Canter 1977), phenomenology (Relph 1976), and structuralism, and Canter's 1997 facet theory of place which integrated with architecture theory.

At the same time, this book raises the question 'how' to explore an abstract social phenomenon, space/place or urban space, rather than examining a particular social phenomenon, Cairo urban space. This has developed two sets of data simultaneously: theoretical data in theories of space/place, and empirical data, in the vignette of Cairo. Hence, to answer the book's question of how to approach Cairo space in order to develop frameworks of place into reading strategies requires a reflective approach with consideration given to both empirical and theoretical data simultaneously.

This discussion is illustrated in Figure T.4, which represents a theoretical reading of this multidisciplinary review of urban space in this book. Initially, architecture and urban theory intrinsically question place-making. On the one hand, philosophy (post-structuralist) questions the nature of space/place, the metaphysical, and the architectural conception of urban. On the other hand, social studies theory explores the social/people experience of urban space, the non-physical space, and the everyday realities of this space. It should be taken into consideration that social studies of place are intrinsically multidisciplinary, involving sociology, psychology, physiology,

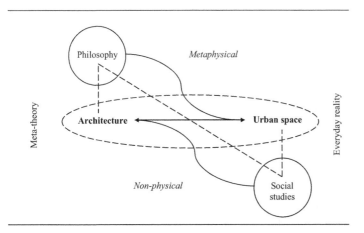

◀ **FIGURE T.4**
Understanding the complexities of the multiple theories

environmental studies, behaviour studies, and architecture (social) theory among others. These multidisciplines require a constant reflective approach in order to avoid the dominance of a specific theory.

Visual thinking

This book is positioned within both a post-structuralist and an architecture background, which thus requires a special emphasis on visual/graphic involvement in data construction, analysis, and interpretation. This involvement varies from the use of photos and architectural drawings to thinking diagrams that manifest and describe the relationship in between architecture and philosophy.

Diagrams are essential attributes of architectural thinking. In architecture, diagrams are representational relational tools in between theory and practice, idea and reality. They involve an extensive vocabulary of symbols, elements and relations, both graphic and verbal with which to present different sets of information (Do and Gross, 2001). Fraser and Henmi (1994: 110) describe how architects 'symbolize . . . intangible factors such as movement, access, sound, view' in a self-conscious design process. Interestingly, diagrams are essentially philosophical attributes as well. Diagrams present a 'temporary moment in between' inside and outside philosophy. They 'outline a process' rather than 'shape a form' (Mullarkey, 2006: 257). Diagrams are presented as problem solvers in philosophy due to their perceptual quality. They are used as representational relation between text and image, word and graph (Mullarkey, 2006: 161).

This book thus involves three types of thinking diagrams. The first is a re-representation, a re-drawing, of the original diagrams provided by the model authors to explain their theory. The second is developed to illustrate the verbal/textual theory of urban space of the model authors when no visual illustration is provided. The third type consists of constructed diagrams which are based on my reading and interpretation of verbal theories as well as presentations of my analytic approach. These visual-thinking diagrams accentuate the interplay between verbal and visual elements of text, in both theory and empirical material. This thus helps the reflection on the development of reading of urban space and particularly the reflexive strategies.

Crossing ~~out~~ through

'A reflexive reading of urban space' seeks the continuous exploration of its 'subject' through this post-structuralist journey. Accordingly, it has developed throughout the course of writing through a process of crossing out. This practice of crossing out, or erasure, is taken from Heidegger and Derrida. Heidegger's crossing out, 'being', 'points to ~~being~~ in its ontological difference from beings' (Young, 2002: 17). Derrida adopted Heidegger's crossing out,

> 'to write a word, cross it out, and then print both word and deletion . . . since word is inaccurate . . . it is crossed out . . . and let both the word and deletion stand because the word was inadequate yet necessary'.
> (Sarup, 1993: 35)

In the context of this book, the crossing out also attempts to replace the ~~crossed out~~ by another synonymous term. The main aim of this book as introduced in this chapter, is to develop a framework to approach urban on two levels. The questions raised in this chapter help to develop and guide the exploration of place through the preliminary reading of the literature on place in theory – philosophy, architecture theory, and social – and Cairo space. In this context, the preliminary reading reflects on the similarities and differences between the involved theories, and a primary framework of interpretation is developed. This primary framework involves a structured exploration of well-defined elements of urban space. 'A turning point' thus considers the involvement of a critical level of interpretation through the deconstruction project that seeks to destabilise both the preliminary reading and framework of urban space. The main aim is thus developed into a reflexive reading of place, adopting a reflexive methodology and crossing out or erasing of the earlier approach, ~~a levelled~~ reading ~~framework~~. Simultaneously, the preliminary reading considers the changing literature of urban space with a view to the production of a trace of synonymous terms that reflect the changing concepts and ideas of urban space. For example, Categories of urban are defined through the trace of ~~elements, set, facet~~, constituents, as will be discussed through this journey.

Through this post-structuralist journey, I propose therefore the development of a not-well-defined framework, which at the same time is not a framework, but rather a set of strategies which neither offers nor limits a structured system to an approach to urban space: a reflexive reading of urban reading. This reflexive reading continues to oscillate between structuralism and post-structuralism, through data construction, interpretation, critical interpretation, and reflections, to another *becoming*.

NOTES

1 'Heliopolis', which means the 'eye of sun'. 'Ain-Shams' in Arabic has been the name of the city since the time of the pharaohs. Today the city is known by both names Heliopolis and Ain-Shams.

2 'An American physicist, Alan Sokal, managed to publish an article in a postmodern journal, *Social Texts*, with the impressive title "Transgressing the Boundaries: Towards a Transformative Hermeneutics of Quantum Gravity" (Sokal, 1996). After publication, Sokal disclosed that the whole article was full of nonsensical but serious-sounding jargon which anyone familiar with physics would have dismissed at a glance as a joke' (Alvesson and Skoldberg, 2000: 182). Although this incident questions the difference between scientific approaches (physics) and social science research, it is still a valid critique of post-structuralism linguistic and textual strategies.

Part I

PRELIMINARY READING OF URBAN SPACE

Chapter 1

APPROACHING A
READING SPACE

Urban space lies in between idea and reality, design concept and people's lives, while simultaneously being embedded within the wider urban context, political, economic, and so on. According to Deleuze and Guattari (1994), architecture as a way of thinking crosses the boundaries of philosophy, science, and art. On the one hand, they consider art and simultaneously architecture as essentially concerned with creation of space experience and sensation (Nilsson, 2004). Architecture and art hence follow a pragmatic approach to 'make space distinct . . . to determine its boundaries', while philosophy and science are more concerned with understanding 'the nature of space' (Tschumi, 1975a: 29–30); while social studies are interested in the experience of people, body, and social in the urban space, both physical and phenomenological, through an empirical approach. This chapter reviews the reading(s) of space/place in philosophy, architecture, and social studies, and reflects on the need to develop an integrated reading of urban space.

> Any study of place [urban space] also entails a bridging of interest across different academic paradigms, particularly . . . cultural studies and human-environment studies.
>
> (Dovey, 2005: 16)

Accordingly, I shall trace the development of associated concepts and definitions of 'space/place' in philosophy, from the classic readings of chóra and topos, to the modern synonymous space and place respectively, as well as the emergent reversion to neo-chóra through Derrida's deconstruction reading of space: 'Khōra'. However, the latter part is discussed in Chapter 3. Consequently, this is complemented by an exploration of social studies of place, which shows a growing interest in the study of the relationship of people to place and attempts to integrate with architecture. Finally, this review investigates the similarities and consistencies between these readings, exploring the prospect of developing a primary framework for reading urban space through five different projections of space: Canter (empirical based, 1977a) and Relph (phenomenological, 1976) in social space; Markus (structuralist, 1980s~) and Tschumi (post-structuralist, 1980s~) in architectural space; and Canter's facet theory of place (structuralist, 1997) integrating social and architectural space.

SPACE/PLACE

In this section I attempt to explore the development of the concept of space/place in philosophy from the time of Plato's chóra, the *Timaeus* (Plato, 1937 [360 BC]), towards the re-emergence of 'neo-chóra', 'Khōra' (Derrida, 1995). I hence adopt and develop Casey's (1997a) classification of these concepts, which I outline as 'chóra/topos', 'space/place', and 'towards chóra'. This classification is represented in Figure 1.1. The first phase, chóra/topos, draws a holistic approach to space/place developed by Plato's chóra and Aristotle's topos, 'Physica' (Aristotle, 1984 [350 BC]). Galileo then presents a transition to modern science and modern space (for further details see Galilei, 2008 [1615]). The second phase considers the modern concept space/place through the works of Newton, Descartes, Locke, Leibniz and reflects the duality between mind vs. matter, rational vs. empirical, absolute vs. relative space. Kant and Spinoza hence mark the transition between modern space and the third phase, towards chóra, the re-emerging interest in chóra and topos, where Kant and Heidegger attempt to reconcile the inherited duality of modern space. The third phase is developed through the works of Nietzsche, Foucault, and Deleuze on the one hand and Derrida on the other, as well as Heidegger. It is Derrida who brought back the concept of space/place to Plato's chóra, through his deconstruction reading of space, 'Khōra'.

▶ **FIGURE 1.1**

Space/place in philosophy and science from Plato to Derrida

Note: The 'Towards neo-chóra' section, based on Agamben and Heller-Roazen (1999: 239) is presented in bold. My additions are presented in both straight and dotted lines.

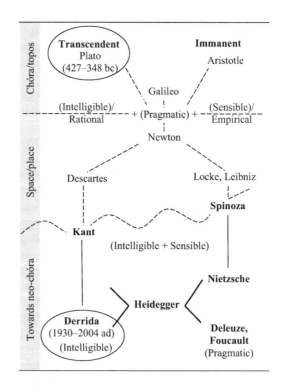

The representation of these concepts in Figure 1.1 is based on Agamben and Heller-Roazen's (1999: 239) diagram, which considered the recent development of the concept of space/place, the third phase. Agamben and Heller-Roazen (1999) draw attention to the relationship between the transcendent, beyond or above the range of normal or physical 'sensible' human experience, and the immanent thinking, 'operating from inside' (Blackburn, 2008) – with Heidegger at the centre. In addition, I also examine Plato's differentiation between intelligible/rational and sensible/empirical ways of thinking; see *The Republic* (Plato, 1892 [380 BC]), as well as pragmatic thinking emphasised by Galileo. The intelligible is 'understood only by the intellect, not by the senses', while the sensible is 'done or chosen in accordance with . . . prudence [good sense]', and the pragmatic is 'dealing with things sensibly and realistically in a way that is based on practical rather than theoretical considerations' (Stevenson, 2010). Finally, Derrida's return to neo-Platonic chóra follows a transcendental rational thinking (Broadbent, 1991b). However, it is important to stress that I am not interested in the differentiation between space and place. Rather my interest is in the exploration of the concepts and themes developed relative to them. Accordingly, 'space' and 'place' are used as originally quoted by their authors, in order not to lose their intended meaning, or space/place is used as a general term. We shall hence review these three phases: chóra/topos, space/place, and towards chóra, whereas Derrida's reading of Plato's chóra, 'Khōra' is discussed in details in Chapter 3.

Chóra/topos

> Chōra and topos were often used synonymously to refer to space and place.
>
> (Rickert, 2007: 254)

Broadbent (1990) reflects on the historic modes of thinking, Greek and Roman, Spanish, and Islamic. These modes constituted three basic ways of thinking: using pure geometric forms, a concern for the experience of human senses, and learning through trial and error. These approaches echo Plato's (427–348 BC) modes of thinking the 'intelligible, sensible and the intermediate' (Launter, 2003: 84). Accordingly, Broadbent (1991b) discusses how the intelligible and the sensible have dominated western philosophy and metaphysics since Plato. Figure 1.2 illustrates these three modes of thinking and their association with the concepts of 'chóra' and 'topos'.

Plato and Aristotle approach the concepts of space/place through 'chóra' and 'topos' respectively. Plato acknowledges the dominance of intelligible chóra, space, whereas Aristotle subverts chóra, as part of sensible topos, place (Casey, 1997a). On the one hand, Plato differentiates between intelligible and sensible ways of thinking, where the intelligible logic is transcendent to and subverts the immanent sensibility (Kymalainen and Lehtinen, 2010; Launter, 2003; Plato, 1892 [380 BC]). This dominance helped to ignore the 'world' between them including pragmatism. Plato also included divine revelation as a mode of inspiration, which considers the 'presence' of God. But he ignored

▶ **FIGURE 1.2**
A reading in ancient philosophy of space/place, chóra/topos

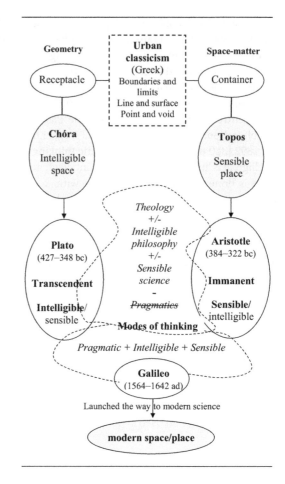

creativity, which relates to works of art, design . . . a world where 'creative people literally "dream up" new, strange and wonderful idea[s]'. Creativity intrinsically challenges Plato's intelligible world and was thus also excluded from his work (Broadbent, 1991b: 37); refer to Figure 1.3. Furthermore, the transcendent intelligible mode designates chóra as a receptacle that receives chaos and conceives order, geometry, and pure forms in particular. This order is thus superimposed on the sensible mode. The receptacle is also intrinsically independent from the properties of the sensible world (Casey, 1997a; Derrida, 1995; Grosz, 1995). I shall elaborate further on the idea of the 'receptacle' through the deconstruction reading of 'chóra' in Chapter 3.

▶ **FIGURE 1.3**
Plato's world (after Broadbent, 1991a: 37)

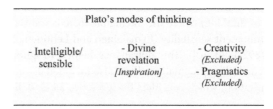

On the other hand, Aristotle (384–322 BC), as the student of Plato, believed in the binary sensible/intelligible. However, he believed in the supremacy of sensible thinking (Broadbent, 1991b; see 'sense and sensibilia' in Aristotle, 1984 [350 BC]). He used a 'logical system', intelligible thinking, to explain his sensible observations of the world (Glusberg, 1991: 58; Tschumi, 1975a: 29). He approached place/space through the concept of space-matter, 'topos', a place which 'subsumed chóra . . . as material [sensible] space' as part of itself (Rickert, 2007: 253; Casey, 1997a). For Aristotle, place is dependent on the 'matter' contained, on their properties, volume, and boundaries which they exchange (Mendell, 1987). Accordingly, chóra loses her reciprocity as it attains the properties of matter and becomes the 'container' space instead of the 'reciprocal'. Accordingly, place is a physical phenomenon shaped by immanent senses rather than a transcendent geometry. Aristotle thus followed an immanent sensible approach to read place as a 'bounding container, the outer limit of the contained matter/body re-joining the inner limit of the containing place' (Casey, 1997a: 58). The container space is defined as a setting 'in which bodies [/matter] are located and move' (McDonough, 2014). The latter view of immanent sensibility could reflect the production of creativity and internal inspiration, which was excluded by Plato. A construction of this reading of chóra, the receptacle and topos, the container is thus presented in Figure 1.4.

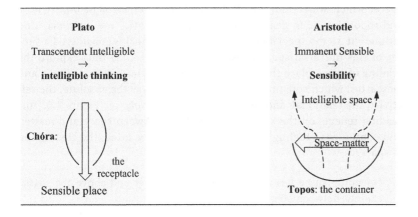

◀ **FIGURE 1.4**
Chóra/topos: intelligible receptacle/sensible container

Subsequently, Galileo (1564–1642) brought back pragmatic thinking along with the sensible when he demonstrated the value of bringing 'pragmatic experiment' together with 'intelligible thinking' and 'sensible observation' rather than accepting the supremacy of the intelligible or sensible ways of thinking (Broadbent, 1991b; Pérez-Gómez, 1994). Accordingly, he launched the way to modern science and eventually to modern space.

Space/place

> Such is modern space . . . not modern spaces. Modern space is ultimately one: . . . absolute, and infinite, homogenous and unitary, regular and striated, isotropic and isometric.
>
> (Casey, 1997a: 193)

The intelligible, sensible, and pragmatic ways of thinking associated with chóra and topos were developed over centuries into coherent philosophies of rationalism: 'reason rather than experience is the foundation of certainty in knowledge'; empiricism: 'all knowledge is based on experience derived from the senses'; and pragmatism: that 'evaluates theories or beliefs in terms of the success of their practical application' respectively (Stevenson, 2010). Simultaneously, many of the early thinkers of space/place, like Newton (1642–1727), Descartes (1596–1650), Leibniz (1646–1716), and Spinoza (1632–1677), worked across both philosophy and science. Philosophy and science are both interested in the study of the nature of space. Philosophy considers the materialisation and production of concepts and notions of reality, while science deals with functions and relationships (Deleuze and Guattari, 1994). Accordingly, interpretations of the 'nature of space' in philosophy and science oscillate between space as 'something subjective with which the mind categorizes things', and space as 'a material thing in which all material things are located', respectively (Tschumi, 1975a: 29).

This duality in the way of thinking has dominated the understanding of space/place for a long time and helped the development of its 'double identity' between abstract and material: mind and matter. Furthermore, it prompted the separation between the container and the receptacle space, which were traditionally brought together through both Chóra and topos. Space, in general, is homogenous, static, and universal, and indifferent 'to the specialness of place'. Place (topos) is considered a subset of this universal space (Casey, 1997a: 137, 133). I thus explore the reading of space/place through the ideas of Newton, Locke, Leibniz, and Descartes, which constitutes four notions of space: the absolute, the relativist, the relational, and the extensive respectively, see Figure 1.5. This reading reflects on the associated duality in between space and matter, and continues to highlight the introduction of the mind, the abstraction of space.

> [Newton's] space has an absolute existence over and above the spatial relations between objects.
>
> (Okasha, 2002: 95)

Newton (1802) while following Galileo's approach of involving pragmatic, intelligible, and sensible ways of thinking, adopted a rational absolute space (Casey, 1997a: 142; Glusberg, 1991). This absolute space 'embraced the void', 'before any occupation by bodies [matter] or forces' (Casey, 1997a: 139, 147). Accordingly, it broke the link in between Aristotle's space-matter: space became detached from the properties and attributes of the inside material. Space is 'an absolute container' that consists of 'determinate boundaries and/or dimensions', which imply that matter is either inside or outside these boundaries (Pries, 2005; Weiss, 2005). This space is also static, whereas matter is mobile between absolute spaces. Consequently, Newton's absolute space is rational, self-sufficient, independent, non-extensive, static and 'immobile': matter is either inside/outside space, and mobile from one static space to another (Casey, 1997a: 142).

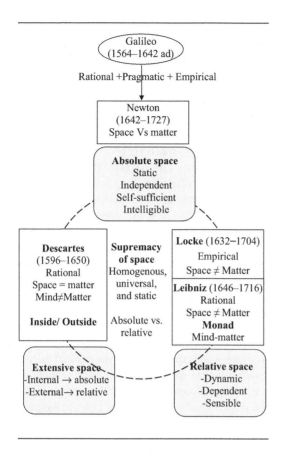

◀ **FIGURE 1.5**
A reading of
Descartes, Locke,
and Leibniz on
space/place, matter,
and mind

> Our idea of Place is nothing else, but such a relative position of anything.
> (Locke, 1975 [1696]: 101)

By contrast, Locke (1975 [1696]) adopted a 'relative space', which is 'a movable dimension . . . of the absolute space it occupies' (Casey, 1997a: 143). He thus 'reasserted the precedence of the senses' through space dependence on sensible relations (Broadbent, 1991b: 40). Simultaneously, he considers space and matter as two separate entities though space is dependent on matter. Relative space is also reduced to a Cartesian position, in which Descartes (1984 [1644]: 43) limits the definition of space to a three-dimension extension 'in length, breadth, and depth'. Locke thus disqualifies the various characteristics of space and matter: temperature, colour, 'size or shape . . . capacity or volume' that cannot 'be converted into calculable distance'. Space is 'nothing but . . . a relative position of anything'. The relative mobility of space is perceived as 'nothing but a change of distance between any two things' (Casey, 1997a: 182, 165; Weiss, 2005). Newton thus defines this relative space as sensible – dependent on its relations to matter and space, and dynamic – mobile, multiple-located, and non-confined to the boundaries of one space (Casey, 1997a: 142, 143).

> We understand that this world, or the universe of material substance, has no limits to its extension . . . we are always able to, not merely to imagine

other indefinitely extended spaces beyond them; but also to clearly perceive that these are as we conceive them to be, and, consequently, that they contain an indefinitely extended material substance.

(Descartes, 1984 [1644]: 49)

Discarding pragmatics, Descartes (2006 [1637]: 29) adopted an exclusive rational way of thinking: 'I am thinking therefore I exist' (Broadbent, 1991b: 40). He attempts to reconcile both theories of absolute and relative space, through the notion of 'extensive space'. He hence presents two spaces an absolute internal space and a relative external place that form the 'extensive space'. 'We sometimes consider the place of a thing as its internal place (as if it were in the thing placed); and sometimes as its external place (as if it were outside this thing)' (Descartes, 1984 [1644]: 46 in Casey, 1997: 157). A key concept for Descartes's (1970) extensive space is the equivalence between space and matter. For him, 'Not only does matter occupy space, but space is matter' (Casey, 1997a: 153) and the extensions of both matter and space are identical. The internal, on the one hand, is equivalent to the volume of matter and hence inherits its physical properties (Grant, 1981; Zepeda, 2009). The external, on the other hand, constitutes the relationships between two matter-spaces. An external 'surface' is 'immediately surrounding' internal matter-space (Zepeda, 2009: 17; Casey, 1997b: 281). In summary, extensive space is a Cartesian space extended in three dimensions that embodies the internal matter-space as well as its relation to other matter-spaces. Zepeda (2009) further differentiates the internal space as a three-dimension volume that comprises matter and the external as a two-dimensional relative 'surface'; refer to Figure 1.6.

▶ **FIGURE 1.6**
Descartes's
extended space

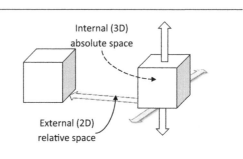

Internal (3D)
absolute space

External (2D)
relative space

Descartes's concept of extensive space is considered as 'ultimately more confusing than clarifying', and his attempt to reconcile absolute and relative space as 'unsatisfying' for adherents of both approaches Casey (1997a: 162; 1997b: 287). Extensive space intrinsically contradicts both concepts of absolute and relativist space, which simultaneously oppose each other. This extensive space also implicates our understanding of space and matter. Both absolute and relativist concepts held the separation between these entities, which were perceived as independent in an absolute perspective and restricted to Cartesian position and relations in a relativist notion (Slowik, 1999). Descartes's notion of an infinite extended space equates space and matter rather than connects them; 'we have the same idea of matter as we have of space' (Casey, 1997a: 153, 184; Descartes, 1970: 184). Extended space co-exists with matter in a dependent association, 'a particular portion of extension that moves with the body [matter] that possesses it' (Schmaltz, 2009: 117). Zepeda (2009) thus questions the potential of this definition of extension to eliminate space, to be replaced by matter as 'there cannot exist a space separate from body [matter]' (Descartes, 1984 [1644]: 16).

Mind/matter

> Space was generally accepted as a cosa-mentale, a sort of all embracing set with subsets such as literary space, ideological space, and psychoanalytical space.
>
> (Tschumi, 1975a: 29–30)

Descartes's notion of extensive space helps to define space as an 'abstraction', a conceptual model of extended matter (Schmaltz, 2009: 115). This conceptual space brings about the discussion on the mind, and its relation to matter. Mind and matter are perceived as two 'distinct' entities, separated on two different planes: the abstract and the material. This separation implicates the connection between mind and matter, although Descartes affirms their interaction (Northrop, 1946; Rozemond, 1995). Furthermore, matter is confined to 'its extension and the various shapes and motions that modify his extension' (Schmaltz, 2009: 115). It 'is strictly passive' dependent on mind 'to move it' (Hattab, 2007: 49). In summary, the mind is abstract, indivisible, independent and non-extensive, whereas matter is material, divisible, dependent on mind, and extensive (Schmaltz, 2009; Rozemond, 1995).

> [For Leibniz] . . . space consists simply of the totality of spatial relations . . . [which] implies that before there were any material . . . space didn't exist.
>
> (Okasha, 2002: 96)

Controversially, Leibniz challenged Newton and Locke's notion of relativist space, particularly space dependence on matter, as well as Descartes's notion of extensive space, particularly the separation between mind and matter and the inferred interaction between them (Northrop, 1946). The presumed duality between mind and matter, space and matter, has enforced a selection between an either rationalist way of thinking created solely in the mind or an empiricist way of thinking dependent only on matter. This dualism also endorsed two sensible spaces: the material and the 'experienced', as well as the conceivable interaction between these spaces, where matter 'causes' the experience (Blackmore, 2005: 251).

For Leibniz, space is not a relative Cartesian position, but rather comprises sensory relations. This 'relational' space is neither matter nor a property or a container of matter. Nevertheless, space did not exist before matter; it is a rational construction of sensory relations. Leibniz thus emphasised the role of the mind that conceives the sensory relations in space (Alexander, 1965; Northrop, 1946; Okasha, 2002). He simultaneously introduced a third entity, a 'middle region': the 'monad' that 'inextricably' connects mind and matter (Casey, 1997a: 180–181). Accordingly, there are three spaces: the material space, the mental space, and a relational space between mental

and material space (Northrop, 1946; Russell, 1945). The latter is the rational construction of the material space, a phenomenal space (Scruton, 2001).
However, these relations are not embedded in the material space but the
monad's transcendent rationalism; the monad is the sole creator of the
material space itself and its inherited relations (Northrop, 1946). The phenomenal space comprises a temporal instance of reality constructed by the
monad (Hamza and Abdelwahab, 2017). And it holds both the primary and
secondary properties of matter as well as the 'spatial, temporal, and causal
relations' (Malisoff, 1940; Northrop, 1946; Scruton, 2001: 530). These properties and relations are confined to the space; they do not exist in either the
material or the mental space. Figure 1.7 thus illustrates the insertion of the
monad in between mind and matter, and the induced in-between spaces,
mental, material, and phenomenal.

▶ **FIGURE 1.7**
Leibniz's monad
inserted between
mind and matter

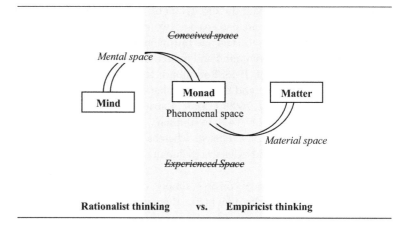

Furthermore, material space constitutes an infinite number of monads
(Scruton, 2001). Each of them has its own 'point of view' of space, a phenomenal space; see Figure 1.8. However, the monad is not located in space nor in
the extended space; there is no spatial relation between them (Casey, 1997a:
179; Russell, 1945). Furthermore, these monads are actively constructing this

▶ **FIGURE 1.8**
The monad's
phenomenal space

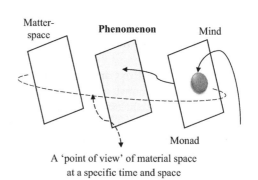

'point of view' of space rather than passively observing it (Northrop, 1946); this construction is transcendent from the mind rather than immanent in the material space. Each point of view personifies 'a minute fraction' in space and time, of the material world, where the monad holds these material properties and a 'loosely . . . spatial position', only temporarily and only on an abstract level in the mind (Casey, 1997a; Furth, 1967; Russell, 1945: 584). Furth (1967: 171–172) thus presents a vivid descriptive image of the experience of the monad in space, which for example is oneself writing. The monad thus 'express' their point of view to oneself; the pen, the hand, and the desk; and 'the interior of the desk', the 'floor beyond', and extend outside the room; the 'others' around you and in adjacent spaces; the 'sky', the earth and the whole 'universe'; and last but not least, time: 'past and future as well as present'. The 'assemblage' of these monads thus creates the material 'relational' space (Russell, 1945: 583).

It is worth noting that the controversy between Leibniz's and Locke's concepts of space has persistently continued. Locke's conservative notion of relative space as a position echoed Newton's concept of absolute space, whereas Leibniz's notion of relational space developed in controversy with Newton's ideas (Casey, 1997a; Northrop, 1946; Okasha, 2002). Newton's account of space is associated with his scientific theories that dominated the seventeenth century after the successful publication of his *Mathematical Principles of Natural Philosophy* in 1678. Consequently, the seventeenth century favoured Locke's relativist space. Subsequently, the disruption of Newton's tradition through Einstein's theory of relativity and quantum mechanics in the twentieth century gave way to Leibniz's understanding of relational space (Casey, 1997a; Northrop, 1946; Okasha, 2002). Nevertheless, the controversy between the two approaches did not resolve, and continues today between the adherents to these theories (Okasha, 2002).

In reflection, the modern urban space, universal, homogenous, and infinite (Casey, 1997a), is a space of matter and body function, where 'form [matter/body] . . . [is] subordinated to function', 'form (ever) follows function' (Sullivan, 1896: 408). Newton's absolute space thus yielded the 'international style . . . all over the world' (Berque 1997: 337). Furthermore, the duality between mind and matter, more particularly the difficulty of their interaction as introduced by Descartes, raised considerable argument about spatiality and urban space (Blackmore, 2005). Consequently, this duality gave way to the revivalism of rationalist and empiricist tendencies in modern urban space (Broadbent, 1990). Rationalists, like Aldo Rossi (1981), prefer working within logic rather than trusting in human senses. Therefore, they are more concerned with the purity of the physical form than with the effect of this design on people's experience. Whereas neo-empiricists, like Gordon Cullen (1961) and Kevin Lynch (1960), trust in human senses. They approach architecture through the people's perspective, drawing on senses, cognition, perception, and so on (Broadbent, 1990).

Towards neo-chóra

The separatist notion of modern space reflects western binary metaphysics; space/place, absolute/relative, space/matter. As discussed, both Descartes and Leibniz attempted to reconcile this separatist notion through theories of interaction and monadology respectively. However, Descartes's attempt was considered 'unsatisfactory', and Leibniz's was discarded until the twentieth century. Interestingly, these binaries were 'constructively and unproblematically combined' through chóra and topos, which brought together relative and absolute space (Casey, 1997a: 136). The study of this era in which many great thinkers have contributed to the development of the discussion, is vast and divergent. Accordingly, Agamben and Heller-Roazen (1999) draw two parallel lines of thought, the transcendent through Kant, Husserl, and Derrida, and the immanent through Spinoza, Nietzsche, Foucault, and Deleuze, with Heidegger at the centre; see Figure 1.1. Significantly, Mullarkey (2006) worked through this diagram to expand and distort, and particularly blur the distinction between the immanence/transcendence binary. Simultaneously, Hillier (2005: 271) reflected on 'Straddling the Post-Structuralist Abyss between Transcendence and Immanence'.

Spinoza's (1632–1677) and Immanuel Kant's thinking (1724–1804) are hence perceived as a transition between this modern space in the seventeenth century reflecting Descartes, Leibniz, and Newton, and the development of the twentieth-century thinking, particularly through Kant's 'phenomenological' approach (Casey, 1997a: 187), and subsequently post-structuralism. Consequently, later approaches to space/place that extend to Derrida [1930–2004] and Deleuze [1925–1995] will consider the reconciliation of these dualities in a re-approach towards 'chóra', and highlighted through Derrida's deconstruction reading of chóra: Khōra. Accordingly, I take special interest in Kant reflecting on phenomenology; and Derrida's deconstruction. The latter shall be discussed further in following chapters.

It is worth mentioning here, Spinoza's theory of monism that considers mind and matter as attributes of a single being. Accordingly, they hold 'a one-to-one correspondence' without the necessity to create an interaction-space or insert a middle region. However, many scholars rejected Spinoza's monism that attributed all matter an immanent mind; a thinking capacity to 'tables and rocks' (Buckingham et al., 2015: 128).

> [T]his body is my body; and the place of that body is at the same time my place.
>
> (Kant in Casey, 1997a: 202)

Conversely, Kant (2008 [1781]) 'skilfully' approached the reconciliation between the simultaneous binaries, rationalist/empiricist thinking, absolute/relative space (Broadbent, 1991b: 40, Casey, 1997a). He believed that either claim to act independently is misleading. A rationalist mode of thinking 'would have nothing to think about' without sensible data, while an empiricist 'would have no capacity to think' without intellect. Both are dependent on each other and accordingly should operate together (Buckingham et al., 2015; Scruton, 2001: 31). He also contemplated the simultaneous 'existence' of internal and external spaces, and emphasised the role of the body in both spaces (Buckingham et al., 2015). I shall call these space-of-body and body-in-space, respectively. The internal space on the one hand,

considers the experiential space-of-body, 'thoughts, feeling' and materiality. On the other hand, the body is the 'medium' to experience the external
space; being inside this experienced space. The attribute of being-in-space is
'responsible for the body's unique contribution to our experience of' space/
place (Casey, 1997a: 214; Buckingham et al., 2015: 170–171). The experience
of the Kantian body-in-space is different from that of Leibniz's 'monad'. The
monad is responsible for the rational construction of a point of view isolated
from the experience, whereas the Kantian body, as matter, exists and experiences material space. Furthermore, the 'perception, memories, thoughts
and feelings' of this body is imposed on the experience (Scruton, 2001). The
body constitutes an internal mental space, and is simultaneously material.
Furthermore, multiple monads exist at different points of view at the same
time, whereas the body's experience is developed through movement in
space and time. 'In being aware of the bodily experience, we must therefore
be aware of the whole spatio-temporal world as mirrored within the bodily
life' (Casey, 1997a: 213).

Accordingly, phenomenology is concerned with the study of the body
experience in place through 'perception, thought, memory, imagination,
emotion, and so on' (Smith, 2009). However, Kant's attempt resulted in the
isolation of internal space from the 'physical world', leaving it trapped in the
'human mind' (Casey, 1997a: 136); 'an ideal internal structure, an a priori
consciousness, an instrument of knowledge', . . . a cosa-mentale (Tschumi,
1975a: 29). Controversially, Heidegger (2002 [1927]) ignored the body; he
was more resolved to approach place through divergent paths rather than
through any one fixed approach. Simultaneously, he introduced Being, in
between the sensible Being (space of solidity, matter which is not the body)
and the where being (space-of-occupation), and hence he emphasised
the mind.

Heidegger's phenomenological study of the body experience in place: the
people relationship to the environment is considered a significant resource
for architects (Seamon, 1982; Seamon, 2000). Heidegger emphasises the
development of personal and social spaces and the implied production of
meaning. He thus argues that the separation between body and place, and
the assumption of directional relations between them is a misrepresentation
of the 'actual lived experience' (Seamon, 2000: 162). Christopher Alexander
and colleagues' (1977) *A Pattern Language* is an 'implicit' study of phenomenology and architecture which relates the contribution of physical design to
the 'sense of place' and thus towards the development of the quality of place
(Seamon, 1982).

Mugerauer (1994) perceives phenomenology as 'midway' between positivism on the one hand and post-structuralism and deconstruction on the
other. Both approaches seek the continuous exploration of space/place, as
phenomena and theory respectively, without prior description, theorisation,
or anticipation, introducing a degree of ambiguity and 'uncertainty'. Furthermore, phenomenology anticipates the clarity and understanding of the
study of place to be uncovered more, and attempts to harmoniously integrate
the relationships between body and place. However, post-structuralism
anticipates 'demonstration of inconsistencies and fragmentations, and the

possibility of multiple interpretations', and intrinsically questions this harmony, order, continuity, and so on striving for change, plurality, and dynamism (Alvesson and Skoldberg, 2000: 152; Mugerauer, 1992; Seamon, 2000).

Concurrently, there is a growing interest in place from various philosophers and thinkers who are trying to locate place in relation to different fields of interest, history, religion, geography, and so on, rather than attempting to find a fixed definition of place (Casey, 1997a). It is worth noting that it was Derrida, Eisenman, and Tschumi who addressed the new emerging philosophy of place through architecture. Derrida's project of deconstruction and the re-emergence of Platonic chóra, Khōra, in architecture will be discussed in the following chapters.

BODY/PEOPLE

> When architects say that people must be taken into account in designing, they are saying almost nothing; they are making vague experiential claim involving themselves.
>
> (Johnson, 1994: 62)

We have traced the development of concepts of space/place in philosophy and science and have thus elaborated how philosophy and science have sought to understand and explain the nature of space/place. They have also tended to focus on concepts of space/place through a rational logic in approaching the phenomenon, understanding and defining its boundaries, which also resonates in architecture. Accordingly, there were two spheres of relations in the urban context, space/place and body/people. However, in general these reflections have emphasised space/place rather than people. This section thus continues to highlight the body/people in urban space through social studies in reflection to architecture and urban theory.

Urban space is provided by the designer, and is continuously used and appropriated by people. Hence, an in-between relationship is constructed through an architectural conception of the ideal urban space, deconstructed and reconstructed through the social realities of people's everyday lives. However, this is not a single relationship at a specific place and time, but multiple relationships through space and time. Simultaneously, architecture has worked in close relationship with social sciences to approach the concept of space/place, highlighting the people in their urban space in particular (Schneekloth, 1987). However, both disciplines have developed a degree of dissatisfaction with the favouring of spatial requirements over people in one discipline, architecture and urban design, and the reverse in the other, social studies[1] (Schneekloth, 1987).

> What architects have begun to realize is that they do have a duty of care towards known users and anonymous others in designing buildings, and that conventional theory, as the discourse of design, intervenes on their behalf while designing.
>
> (Johnson, 1994: 68)

Moore explored the possible integration of social studies with architecture theory. However, he considered architecture as a normative theory with design guidelines and rules; and environmental studies as positivist explanatory theory, 'answering the why' (Moore, 1997: 4). Accordingly, he considers three possible links between social studies and architecture theory. The first examines the development of design guidelines and looks at post-occupancy evaluation studies. The second considers the development and testing of architecture theory based on social studies. Groat (1984) reflects on Relph's (1976) phenomenological study of place and in turn Seamon (2000) reflects on the significance of Relph's approach to a 'sense of place' that goes beyond the perceived place. Groat (1995b) and Sime (1985) drew attention to the similarities between Relph's approach and Canter's (1977a) in developing similar three-part models of place though each was developed within different disciplines and epistemological backgrounds: human geography and environmental psychology respectively. Both are introduced later in the following section and discussed in details in Chapter 5. Groat (1995b), Sime (1985), Groat and Després (1991), and Lawrence (1982) thus explored the involvement of these models, either directly or indirectly, in earlier approaches to social studies. However, they also criticised them for the limited attention paid to the physical attributes of urban space dealt with by the architect and urban designer, and to people's interaction with these attributes. Subsequently, Groat and Després considered the potential of integration with architecture theories, particularly Canter's model, through questioning architectural style, composition, type, urban morphology, and place.

The third linkage attempts to integrate social studies and architecture theory 'at a conceptual level' through a perspective/normative theory, 'making strong normative statements . . . and explaining why [social behaviour and] relationships . . . occur' (Moore, 1997: 30). Moore (1997) introduces Robert Venturi and Christopher Alexander among others as key figures in the development of this approach. In his book *Complexity and Contradiction in Architecture*, Venturi (1988) puts the emphasis more on architecture design, 'the richness and ambiguity of the modern experience' (Moore, 1997: 31). He expresses his ideas in two ways, 'perception' and 'program and structure'. Venturi then extends this architectural approach to the urban scale, landscape and townscape, where contradiction between the inside and the outside is an essential characteristic of urban architecture (Broadbent, 1990). Christopher Alexander et al. (1977) is more controversial. Moore (1997) on the one hand considers Alexander's approach through normative theories as harmonious to both architecture design and environmental-behaviour studies. On the other hand, Sime argues that 'Pattern Language' is explicit evidence of the dangers of 'ignoring the relationship people have to places' especially as outlined in the early works of 'Pattern Language' and 'Timeless Way of Thinking' (Sime, 1985: 31).

Lawrence questions the involvement of structural theories of place, which eventually opened the way to post-structuralist involvement led by Derrida. Accordingly, he discusses the application of these approaches in architecture through 'the syntax of the built environment', and 'architecture semiotics' (Lawrence, 1989: 51, 57). Moore criticises Lawrence's approach as rich in theory, yet not 'so elegant, diagramable or memorable' (Moore, 1997: 43). Another structuralist approach to place can also be traced in Canter's facet theory of place, which sought to integrate with architecture through Thomas Markus's (1982a) theory (Canter, 1997). This theory is also discussed further in the next section.

> If postmodernity is to be the age of form (modernity have been that of substance and function).
>
> (Berque, 1997: 337)

Social studies have worked through an empirical and pragmatic approach (Broadbent, 1988; Seidman and Alexander, 2000). Accordingly, these studies were interested in the phenomenological experience of urban space, placing the body in a comfort zone; 'one should design for visual, thermal, and aural comfort and for delight' (Broadbent, 1988: 74). However, the late '80s witnessed a major change in direction in social studies, which embraced post-structuralism. Accordingly, social studies of place converted from a foundationalist and humanist position into a post-foundationalist and post-humanist one (Schneekloth, 1987).

Foundationalism is basically concerned with setting the foundations, the scientific methods, and categories of an approach to a unified social truth (Seidman and Alexander, 2000). Post-foundationalism recognises a different reality, 'a more complex, multidimensional type of argumentation . . . that moves between analytic reasoning, empirical data, normative clarification and remains reflective about its own practical social implications' (Seidman and Alexander, 2000: 2). Accordingly, a post-foundationalist approach to social studies considers general theories, 'without foundation', like post-structuralism and cultural studies. Hence it blurs the division between the different disciplines and approaches to social theory, sociology, environmental psychology, behaviour, and urban studies (Seidman and Alexander, 2000). Accordingly, in the following section we shall explore this post-structuralist context in social and architecture theory, particularly in reflection to the body/people in urban space.

Consequently, a humanist approach asserts the transcendence of humanity, that is the body/people, which is reflected in the appropriation of architectural space to people's needs and requirements (Macarthur, 1993). As discussed in the previous section, the 'body-in-space' idea was initiated in the debate through Kant, Heidegger, and phenomenological reflections and concerns. This was reflected through an interest in experience, body activities and movement on the one hand, and the experience of the senses, mind-body relation on the other. Post-humanism therefore considers a different notion of humanism, which draws on the materiality of the body through its inscription in context, space, and time (Macarthur, 1993). Simultaneously, the involvement of the context initially reflected the instant physical setting of place with special concern for historic settings. However, this context was further developed to include the landscape, and the urban context, social, political, economic, and cultural.

Post-structuralist architecture took this position further by detaching architecture space from social and functional requirements (Macarthur, 1993). Berque (1997: 336–342) traces the evolution of modern space into post-structuralism through an illustration of art, architecture and urban. He argues that post-structuralism has 'led us beyond any modern or non-modern, western or non-western culture'. On the one hand, modern space subverted form through functionalism. Post-structuralist space, on the other hand, is a space of form dissociated from function. However, it risks the conversion of functionalism into formalism, which takes after 'past or foreign styles'. Simultaneously, modern space also, subverted the body to function, 'to the accomplishment of series tasks'. Berque (1997) brings about the 'form of the body' along the 'form of space' as an attribute of postmodern/post-structuralist space. Accordingly, the duality of modern space as discussed in the previous section, which helped to distance realities of space that is 'architecture presence', from its representation, was submersed in postmodern space. The later therefore constitutes the 'landscape', the wider context of place, form, body, and space.

PROJECTIONS OF URBAN SPACE

> The complex cultural, social, and philosophical demands developed slowly over centuries have made architecture a form of knowledge in and of itself.
>
> (Tschumi, 1980: 102)

In reflection to this reading, urban space is perceived as a medium of thoughts, actions, and activities; interactions, between the self and the other as well as between people and their spatial environments. These interactions imply a rational and/or emotional relation(s) between their components. This understanding of urban space is not limited to the physical nor the social space; it is dynamic, changing, and constitutes multiple relations and processes (Nilsson, 2004). 'Urban space' is concerned with both the physical and social as well as the conceptual 'space of knowledge' (Tschumi, 1980). Pérez-Gómez (1994: 2) considers architecture and urban design as 'embodying wisdom' within the built environment. It is intrinsically appropriate now to explore previous models, frameworks, and approaches to space/place in architecture and social studies.

To begin with, we shall explore the early three-fold models of Edward Relph (1976) and David Canter (1977a), which were developed through social studies, as well as repetitively mentioned in the various studies of urban space. Furthermore, many authors sought to integrate with architecture theory (Groat, 1995a; Groat and Després, 1991; Sime, 1985). Subsequently, we shall also investigate the 'facet theory of place' developed by Canter (1997) to integrate his early model with architecture through Thomas Markus's (1980) theory. The latter draws on the famous Vitruvius triad of architecture space, and is intrinsically significant to include in this reading. Controversially, Bernard Tschumi (1981a) challenges the dominance of

Vitruvius reading of architecture/urban space, and thus introduces his post-structuralist three-fold approach. Finally, this review investigates the similarities and consistencies between these readings, exploring the prospect of developing a primary framework for reading place through five different projections of space: Canter (empirical based, 1977) and Relph (phenomenological, 1976) in social space; Markus (structuralist, 1980s~) and Tschumi (post-structuralist, 1980s~) in architecture space; and Canter's (1997) facet theory of place (structuralist) integrating social and architecture space; see Figure 1.9.

▶ **FIGURE 1.9**
Five projections
of urban space in
theory

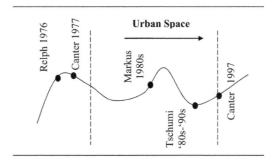

Accordingly, I shall attempt to develop a brief reading of these theories in order to identify the similarities and dissimilarities, through the identification of the involved elements/components of urban space and any apparent relations between them. Accordingly, these theories are generalised through the primary framework of urban space. This preliminary reading is followed up in Chapters 5 and 6, by a thorough reading. However, it is important to emphasise here, that the main aim of this preliminary reading is to identify the constituents of urban space, in general, as well as the existing relations between them, which recognises these different approaches rather than neutralising them.

> [The multidisciplinary study of urban space] does not entail collapsing them into a newly totalizing metanarratives. Rather, it is recognition that different knowledges, soundly based within their own paradigms may be useful to a multiplicitous understanding of the built form.
>
> (Dovey, 2005: 16)

Social space

> [D]espite the vastly different analyses presented by Canter and Relph, both authors actually propose three-part models of place that are described in very similar terms.
>
> (Groat, 1995b: 3)

Canter and Relph presented similar three-part models of place. Relph considered the experience, the 'sense of place'; place is represented as physical features, activities and functions, and meaning (Relph, 1976). Canter's cognitive approach simultaneously read place as physical environment, actions, and conceptions (Canter, 1977a). The two models introduce three inter-related elements of place that are irreducible to one another, and related through union and intersection respectively.

On the margin, the two models reflect through the reading of architecture and philosophy in this chapter that constituted mind, body/body and space/place. I hence attempt to generalise both models through an examination of their similarities. Hence these models are abstracted into place (physical attributes and features: space/place), people (actions and activities: body), and people-place (meanings and conceptions given by people to place: mind) as well as the relations (R) between them, developed here through unions and intersections (Figure 1.10).

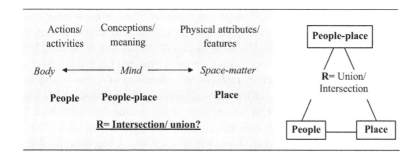

◀ **FIGURE 1.10**
Abstraction of Relph and Canter models

The aim of this book is to develop a general model of urban space that could help in understanding the different readings; for example, the 'people' could refer to the movement of body, architecture programme, people's activities, space function. Simultaneously, this abstract reading approaches the content of place, rather than the wider context, although 'people', for example, could also refer to the social context. This place is then projected on to the other reading.

Architectural space

For Markus, architectural space involves function (user activity programme), form (style and geometry) and space (place morphology and organisation). These three entities are 'in principle, independent of each other'. However, there is a 'typical conjunction of function, form and space' that forms a clear relationship between them. This relation is a social product; that is architecture is carrying meaning to society in general and users in particular (Markus, 1986: 26–27).

Markus's reading is projected into the abstract reading of both Canter and Relph's models. Markus's reading is hence read as place (space-and-form), people (function), and place-people. The latter considers the main difference between the two abstractions. For Markus, place carries meaning to people rather than people giving meaning to place, as previously introduced by Relph and Canter. Simultaneously, Markus does not consider the mind-body relation in his reading; meaning is immanent in place and people are reflected through space of function (Figure 1.11).

▶ **FIGURE 1.11**
Markus's model of
reading of urban
space

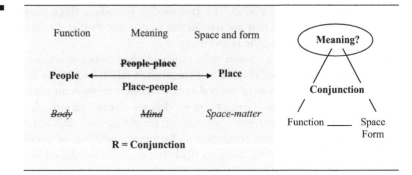

Tschumi's (1981a, 2001: 255) approach consists of two complementary trilogies. In the first, architecture space is 'the combinations of spaces, events, and movements without any hierarchy or precedence among these concepts'. Simultaneously, he refuses any relation between expected place and expected use. Consequently, disjunction replaces the relation(s) and non-relation(s) between these mutually exclusive entities, body movement, space, and event (Tschumi, 2001). The second trilogy approaches a different dimension of architecture space, through 'concept, context, and content' (Tschumi, 2004). These two trilogies are projected into the abstract model. Significantly, there is no people-place relation, no meaning; place content constitutes the experience of the movement-of-body-in-space, which brings about the event (see Figure 1.12). However, the event goes beyond the abstract place. In Figure 1.13, on the other hand, there are no people; the concept-form relation is emphasised, while the context goes beyond the abstract model.

▶ **FIGURE 1.12**
Tschumi's first
trilogy

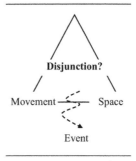

▶ **FIGURE 1.13**
Tschumi's second
trilogy

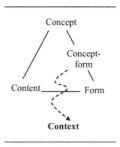

Tschumi's reading disregards the people-place or place-people relationship (Figure 1.14). It consists of two readings that separate space-matter. The first accentuates the intimate presence of body-in-space, and the second considers both the abstraction and realisation of the form through the 'concept-form'. Accordingly, the two readings help to accentuate the disjunction between people and place. Simultaneously, the previous approaches of Canter, Relph, and Markus constitute three irreducible, interlocking, and well-defined elements of place. Tschumi on the other hand, blurs the distinction between place constituents through the introduction of intermediate, in-between definitions, that is body-in-space and concept-form. Therefore, the anticipated relations in the proposed abstraction are weakened and destabilised through the concept of disjunction. The involvement of 'meaning' is questioned as well as the introduction of the 'event', and the 'context'. For Tschumi, an event is an occurrence, an incident that could be in conflict with either/both place and people. Accordingly, the involvement of the event in the abstraction is questioned. However, it needs to be taken into consideration that the event should not be taken as an equivalent of meaning/conception of place nor of the context.

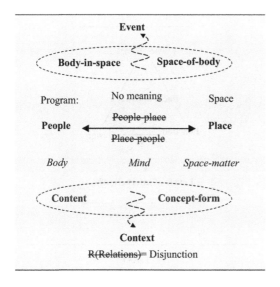

◀ **FIGURE 1.14**
Tschumi's model of a reading of place

A socio-architectural space

> Canter (1997) has developed a more complex 'facet theory', suggesting four interrelated facets of place: functional differentiation, place objectives, scale of interaction, and aspects of design, each with a number of subcategories.
>
> (Gustafson, 2001: 6)

As discussed earlier, Canter developed his earlier model (1977a), into the 'facet theory of place', which integrated architecture space through Markus's trilogy. The facet theory of place provides a structural framework to address the multiplicities and dynamics of social space (Canter, 1983a; Canter, 1997). Another important contribution of this theory is the concept of 'facet', which is a unit of categorisation, a set of elements rather than a single element as in his earlier model (Canter, 1983a).

Accordingly, not only does it help the introduction of the constituents of place but it also elaborates further on them through the definition of their included subsets. The facet theory of place consists of four facets (Figure 1.15):

*   Facet A: functional differentiation, which corresponds to and develops the activities and actions, people
*   Facet B: place objectives, which extends the people-place relation, the conception
*   Facet C: level of interaction; this facet goes beyond the abstract place, as it reflects on levels of contextual interaction
*   Facet D: aspects of design, which develops the notion of place through the integration of Markus's trilogy.

(Canter, 1997; Gustafson, 2001)

▶ **FIGURE 1.15**
Canter's later approach to urban space (1997)

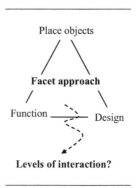

▶ **FIGURE 1.16**
Reading the 'facet theory of place'

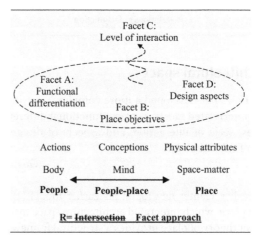

Canter's facet theory of place resembles some of the features included in Tschumi's reading of place. Both approaches extend to the context. Canter considers the level/scale of the context – local, intermediate, distant – whereas Tschumi extends to the wider urban context, historic, cultural, and so on. Also, both readings blur the boundaries of the well-defined elements of place, in favour of sets of elements and intermediate sets, respectively. Consequently, meaning is deleted from Tschumi's reading. However, although this is not discussed explicitly by Canter, it is embedded in the relations between the facets.

Primary framework of urban space

To this end, a preliminary reading of five theories of space/place is introduced: Canter and Relph (1970s), Markus and Tschumi (1980s), and Canter again (1990s). This preliminary reading helps the development of a primary framework of urban space comprising people, place, and people-place (Figure 1.17). This framework provides a general approach, which assists the recognition of the similarities and differences between these theories. Consequently, Markus's approach provides a different reading of people-place relation through place-people, with a different understanding of meaning in place. The abstract model considers the production of meaning through people, while Markus considers the production of meaning through architecture. Tschumi however, rejects people-place relation, and simultaneously, meaning. Canter (1997) on other hand, embeds meaning in the relation between people, place, people-pace as well as context. Accordingly, this preliminary reading helps to question people-place relation(s) and meaning. In addition, Tschumi introduces the production of the event, through the body experience in place; however, this 'event' is not discussed in other theories though they do reflect on the experience of urban space. Finally, both Canter (1997) and Tschumi consider the context, as well the different levels of approaching space/place. The relations (R) between the constituents of urban space have been highly controversial, through intersection, union, conjunction, disjunction, and facet approach. Accordingly, the primary framework of place is developed through people, place, and people-place relations. This model questions meaning, event, and place interrelation (R).

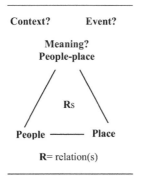

FIGURE 1.17
Primary framework of urban space

Finally, this framework acknowledges the insertion of space/place in context as well as the temporality of reading frames or levels of urban space. The aim is to explore this context and levels, as well as their relationship to urban space.

INTEGRATIVE READING OF URBAN SPACE

> Remember: architecture was first the art of measure, of proportions. It once allowed whole civilizations to measure time and space. But speed and the telecommunications of images have altered that old role of architecture. Speed expands time by contracting space; it negates the notion of physical dimension . . .
> Of course, physical environment still exists.
>
> (Tschumi, 1989: 216)

This chapter has reviewed the development of the philosophy of space/place; of chóra/topos, in ancient philosophy, which considered space/place as whole. This holism was bisected by modern dualism, which reflected on internal and external spaces, matter and space, contained and container. In an increasing affinity towards ancient philosophy and attempts to recover its 'holism', 'modern space' developed a concern with mind, as intermediate between space and matter, as well as a phenomenological emphasis on the body in space. These considerations developed into the twentieth-century approach that thinks in terms of heterogeneity, indirections, complexity, and dynamics of place, and reverts to Platonic chóra, space, mind, and body. Simultaneously, we reviewed the development of architecture space from Vitruvius to Derrida, through a reading of Markus and Tschumi. Thus, the reading of architecture space is explored through both a minimalist and a maximalist approach. The main aim of this book, as developed in the introduction, is to develop a framework to read place on two levels, place content and context, which are endorsed through these two approaches. However, we have also made it clear that these approaches, though they complement each other, do not operate on two levels. Accordingly, the imaginary boundaries of classification between content and context, inside and outside every space, physical, corporeal do not exist, as well as levels or layers. On the contrary these spaces, mind, space, body, conceptions of space and body, perception and experience in space and context, interact and interrelate in complex and dynamic ways. Consequently, this review was complemented by an exploration of the social studies of place, which showed a growing interest in the study of the relationship of people to place through an empirical approach. Social studies of place, like philosophy and architecture theory, recognise the heterogeneity, multiplicities, and dynamics of place, as well as the dissolving boundaries between the different disciplines. Both social studies and architecture theory have thus acknowledged the need to integrate their approaches to place. These integrative attempts have helped to develop two models of place, by Relph (1976) and Canter (1977a), which have influenced many social studies. In addition, Canter developed his model further in order to integrate architecture theory. Accordingly, this chapter attempts a preliminary reading of these models of place together with both Markus's and Tschumi's theories of architecture space. Finally, this review provides an overall reflection on space/place to assist in the development a primary framework of place, which will be elaborated later.

NOTE

1 We are using a general term 'social studies'; Moore (1997) uses environmental-behaviour studies as a general term to environmental psychology, behavioural and social geography, environmental sociology, human factors, social and behavioural factors in architecture, and urban social planning.

Chapter 2

ARCHÉ-DECONSTRUCTION[1]

Tschumi asked Derrida to collaborate with Eisenman in making 'Parc de La Villette' in Paris in 1985. Derrida promptly perceived two spaces, architecture and writing; architecture is 'a writing of space' (Derrida, 1986). Derrida's proposition set off an awkward interaction between architecture and deconstruction, and helped to raise many questions about architecture and deconstruction; how do they interact? Controversially, Derrida was initially sceptical 'about the possibility of any connection between the . . . reality of architecture and the free play of deconstruction'; however he came to the realisation that 'on the contrary, the most efficient way of putting deconstruction to work was by going through art and architecture' (Soltan, 1996: 238). But what is there to deconstruct in architecture anyway? And why would anyone want to do it? (Broadbent, 1991c: 63).

> Who needs Derrida anyway?
>
> (Broadbent, 1991a: 30).

Interestingly, a dialogue between architecture and deconstruction has already started without any association with Derrida and his deconstruction project. Johnson and Wigley (1988) associated the work of this movement with 'constructivism' and thus called it 'deconstructivist architecture'. Constructivism is a Russian movement that started in 1913 to the 1940s, and involved experimentation and 'abstraction' of modern geometric forms (The Art History Archive). Deconstructivist architecture thus involved the work of Bernard Tschumi, Peter Eisenman, Frank O. Gehry, Zaha Hadid, Daniel Libeskind, Rem Koolhaas, and Coop Himmelblau, who re-investigated and questioned modern architecture in a 'similar process [to constructivism that] it is hardly surprising that they discover forms much like those of constructivists' (Broadbent, 1991a: 24). Deconstructivist architecture is thus identified as a 'performative act' of 'desire, displacement, dislocation, subversion' of forms (Soltan, 1996: 269; Wigley, 1995). It is a formalist approach that rejects the modern tradition of pure forms; as well as the transcendence of philosophy to architecture practice. Architecture, 'free from the influence of any language or philosophy' works from with itself to deconstruct the 'inherited' structures and forms (Broadbent, 1991a: 25; Culler, 2003b).

Broadbent (1991a: 11) highlights a parallel perspective, 'architecture deconstruction', which relates architecture and Derrida's deconstruction. 'Architecture deconstruction' differs from 'deconstructivist architecture', not only through the in/exclusion of the work of Derrida respectively, but also the approach to architecture. Deconstruction brings about a shift from an interest in the creation of new images to 'creating a new concept of architecture space. It is interested in the architecture way of thinking, 'theory . . . and programme', and particularly the relation between architecture and philosophy, and thus emphasises 'the deconstruction of writings on architecture . . . from say Vitruvius . . . onwards' (Broadbent, 1991c: 63, 92; Glusberg, 1991: 9).

Following this discussion, this book is interested in the study of deconstruction in relation to architecture as a way of thinking, with emphasis on the concepts of space/place. Furthermore, the study of deconstruction strategies in the next section will highlight that deconstruction operates from within the discourse rather than applying from outside. Accordingly, it is evident that deconstruction is not perceived as a transcendent philosophy to be applied to architecture theory and practice, but rather an intertwined process concerned with the understanding of the concepts of space/place. The second part thus highlights the displacement in theory of architecture, formerly dependent on the writings of Vitruvius, bringing about new ideas and concepts of space as discussed by post-structuralist authors, particularly Bernard Tschumi. The chapter thus concludes by introducing three deconstruction-reading strategies that operate within architecture, to approach the deconstruction reading of the concepts space/place; a deconstruction reading of space/place, Plato's chóra, and Cairo urban space is thus presented in the following chapter.

WHAT IS DECONSTRUCTION?

> What deconstruction is not? Everything of course!
> What is deconstruction? Nothing of course.
>
> (Derrida, 1991a: 275)

Many authors, including Derrida, have attempted to provide a definition of deconstruction. These attempts have mused on the impossibility of developing a clear definition of the word, of identifying what is deconstruction and what is not. I am not interested in investigating the meaning of deconstruction, but rather in exploring the potentials that exist within this project to approach the reading of urban space.

> Deconstruction is neither an analysis nor a critique . . . is not a method and cannot be transformed into one.
>
> (Derrida, 1991a: 273)

Deconstruction is not a method, 'pas de méthode', a critique, an analysis, or a reading; there are no steps, rules, or criteria to be applied (Bennington,

2001; Culler, 2003a; Derrida, 1991a; Lucy, 2004; McQuillan, 2001; Royle, 2000). McQuillan (2001: 5) emphasised the mistranslation of 'pas de méthode', which implies that deconstruction is not a method and simultaneously 'a step in, or towards, a methodology'; that is 'an impossible method'. The 'impossible' here is defined as opposite to the 'potential' and not the 'possible' (Kessel, 2007). We shall thus apply this definition to all the impossible clauses in deconstruction, impossible method, impossible place, and so on. Deconstruction is not a method of 'reading or interpretation' that is applied from outside the discourse; deconstruction takes place within (Lucy, 2004). It is not a method that searches for a unity in meaning but appreciates the singularity of each reading (Bennington, 2003). In other words, deconstruction does not present a 'systematic and closed' procedure for reaching the meaning (Royle, 2000). Culler (1982: 85) describes deconstruction as a strategy rather than a method, 'a philosophical strategy . . . a strategy within philosophy, a strategy for dealing with strategy'. This book adopts Culler's approach, which takes the art of planning rather than the systematic application of specified techniques (Pearsall, 2001). Accordingly, deconstruction is considered as a set of strategies that are developed and approached from within the discourse of urban space presented through this book, without hierarchy or precedent arrangement. These strategies acknowledge the power of the author and the associated reading of urban space. They do not aim to develop a coherent piece of writing which reflects one truth and one meaning. These strategies will be explored through the following points as well as through the development of the book itself. Four strategies of McQuillan (2001) are introduced, adopted and developed here: 'deconstruction in context', 'a history without history', 'deconstructing binary oppositions', and 'embracing the margin', and I shall add two more, induced from these categories and highlighted owing to their strong relevance to the context of this book: the event and the meaning.

> There is no set of rules, no criteria, no procedure, no programme, no sequence of steps, no theory to be followed in deconstruction. . . . Once we have overcome this naïve desire for a formula to academic socioeconomic success, and opened ourselves up to the possibility of another way of thinking about the act of reading, then we can begin to orient ourselves towards the questions raised by deconstruction.
>
> (McQuillan, 2001: 4)

In context: Il n'ya pas de hors-texte

'There is nothing outside of the text'. This statement is considered by McQuillan (2001: 35) to be the most misunderstood piece of Derrida's work. The common interpretation implies the alienation of the deconstruction discourse from the wider social, political, historical context, 'and reality all together' (McQuillan, 2001: 36). Bennington and Derrida (1993: 85) illustrate how this understanding has developed the 'historian's objection to Derrida' through the claim to return to the context in order to understand the text. They question the validity of any reading outside the context, such as a quotation, which is in reality inserted into a new context.

According to McQuillan (2001), il n'ya pas de hors-texte rejects a signi-fied transcendent, which is outside the text, that is separated, 'a reality that is metaphysical, historical, psycho-biographical etc.' (Derrida, 1997b: 158). This rejection holds two implications for the content/context relationship and the text/author relationship.

The content/context relationship may reflect on what Lucy (2004) describes as 'inside/outside transcendental difference', where the text is the inside, and the context is the outside, for example, politics, social, and so on. (Bennington and Derrida, 1993). However, an intrinsic position of Derrida and deconstruction is the refusal of distinct separation between a binary opposition in favour of a blurry boundary and de-limitation; 'drawing a line in order to say this is the text [content] and this is the real world [context] to which it refers to is a false distinction' (McQuillan, 2001: 37). This statement therefore, works on blurring the separation between the text and context through:

• The recognition of the inscription of the context, social, historical, and so on, inside the content, 'we can access them [in the text]' (McQuillan, 2001: 38).
• Accordingly, the text, which is defined beyond written words and char-acteristics, represents a trace of contextual reality, 'which gives the text its meaning' (McQuillan, 2001: 38). The 'trace' is the factor of blurriness in between content/context, through its continuous oscillation – non-decidability – between presence and absence (Collins et al., 2005), i.e. context/text relation and non-relation.

Finally, Bennington and Derrida (1993: 90) question the historian's notion of putting the text back into context, when no one can actually re-build 'a context'. The text is not context free; accordingly, a new context and a new textual trace are built 'to the extent that every trace is the trace of a trace'.

In the same sequence, the distinction between text/author, which is tra-ditionally separated through a false transcendental outsideness, is blurred (McQuillan, 2001). Derrida considers the text/author relationship as sup-plementary[2]; the author's life is inscribed in the text which becomes a part of the author's life, consequently shaping and re-arranging the author's life after the text (McQuillan, 2001).

It is worth noting that the blurriness between content/context and text/author does not imply a homogenous production of truth, but rather helps to embed and recognise the multiplicities within the text that are inscribed in the context (Bennington and Derrida, 1993; McQuillan, 2001). This also helps in understanding the 'temporality and singularity' of text which is simultaneously a referent of an outside context. Accordingly, the content is caught in another context of reading 'non-identical with authorial inten-tions', and hence, the text is subjected to 'misinterpretations and misappro-priation' and is continuously changing meaning and intention through time and context (McQuillan, 2001: 36; Norris, 2004).

A history without history

> If deconstruction historicises, it does so by means of a 'history with-
> out history'. That is to say an idea of the historical which has to be
> thought outside the logocentric conceptual schema which surround
> the traditional use of the term 'history'. In this way deconstruction is
> both historical and cannot be assimilated to an easy historicism.
>
> (McQuillan, 2001: 35)

In this section McQuillan (2001: 35) exposes another 'myth about deconstruction that it is ahistorical'. He reflects on how 'historicism' has dominated
'material criticism' through Marxism, cultural materialism or neo-historicism
for example. This brings about the case of Cairo's representation discussed
in 'the threshold' as an example of this dominance. It was demonstrated that
the representation of Cairo was dominated by a singular historicist socio-
religious reading. McQuillan (2001) considers the deficiency of this type of
reading which calls attention to history through both the production and
explanation of the discourse. This brings history to a state of 'history beyond
history', a state that entails an approach to history without understanding the
concept of history. Accordingly, McQuillan (2001: 35) discusses the historicity of deconstruction through its reading of the 'non-identity' of the 'present',
a present which is always in relation, tracing the past or the future 'which is
nonrecoverable, not even potentially so'. Accordingly, he explains the formulation of the meaning of text through history.

> The moment we use the term [word] . . . we enter into a discourse which
> is already established and which has created a certain meaning for the
> term . . . which we cannot help but to be implicated in even if we wish to
> reject this use of the word.
>
> (McQuillan, 2001: 32)

This discussion of the historicity of deconstruction reflects on Derrida's famous
quotation the 'text is never innocent' (Gannon and Davies, 2012: 78). Each text/
term has a history, a long process, where the term continued to change. However, the singularity of each term is expressed in relation to 'a specific context'. Accordingly, the variety of 'specific context(s)' through time and location
expresses the plurality of each term. A significant notion of this process is that
it hides 'the tracks of this historical formation, not allowing the concept to be
read as historically' (McQuillan, 2001: 30). Accordingly, the origin, the logocentric, of the term could not be identified. This discussion of the historicity
of deconstruction helps to bring up a discussion on the event as a temporal
historical instance, which will be further explored in the following section.

An event

> Deconstruction takes place, it is an event that does not await the
> deliberation, consciousness, or organization of a subject, or even of
> modernity. It deconstructs itself. It can be deconstructed.
>
> (Derrida, 1991a: 274)

Through Derrida's persistent questioning of the concepts and categories of metaphysics, he questions the concept of the event outside the binary oppositions of presence and absence, reality and actuality (Lucy, 2004). Lucy demonstrated that Derrida's interest in the event, among other metaphysical categories, is a manifestation of his resistance to a future dominated by traditionalism/historicism; 'the tendency to circumscribe and confine and limit, to determine the limit, the range of what may be asked and what may not, to what may be believed and what may not' through models from history/tradition' (Berlin, 1969 in Lucy, 2004: 37). This is a phenomena that has dominated the twentieth century (Lucy, 2004), as well as the contemporary representation of the city of Cairo. Thus, we shall try to understand Derrida's event in association with the metaphysics of event:

> An event cannot be reduced to the fact of something happening. . . . If I am sure that something will happen, then it will not be an event. . . . It is what may fail to come to pass. . . . So there is no event without surprise.
>
> (Derrida, 1994: 254–255)

In metaphysics, the 'event' is regarded as a controversial category; its apparent simple definition in philosophy as 'something that happens' is not that simple; for example it re-questions the definition of the term 'happen' (Casati and Varzi, 2008). The metaphysical event is considered in close relation to perception, action, language, and thought. However, the integration and/or separation of these four categories is not well defined. Simultaneously, the definition of the event is also considered in a comparative/complementary relationship to other metaphysical categories of objects, time, facts, and properties (Casati and Varzi, 2008). Furthermore the event is defined through 'spatio-temporal specificity'; 'the metaphysics of presence' (Lucy, 2004: 33). This presence is complemented by the 'significance of the event', the 'event-ness', which is not simply present in the boundaries of space and time (Lucy, 2004: 33). This reading of the event in metaphysics is represented in Figure 2.1.

> The event must be considered in terms of the 'come hither', not conversely. 'Come' is said to another, to others who are not yet defined as persons, as subjects, as equals.
>
> (Derrida, 1994: 253)

The dual presence of the metaphysical event in time and space implies a fusion of 'observable action' (Kessel, 2007). In this sense, events are described as things rather than interpreted in the text. Events happen 'outside the text'; they are not made (Lucy, 2004). Controversially, Derrida's event does not exist (Kessel, 2007), but equally, it is not virtual (Derrida, 1993). The event 'set free from metaphysical' constraints is set free to become, to happen. The event is interpreted inside the text and implies a fusion between the event categories, perception, action, linguistics, and thought. 'To see an event is also to make it'; this also involves thought, the decision to see, to make the event (Lucy, 2004: 36). Furthermore, a metaphysical event is also an isolated moment in time which belongs to 'historical temporality' and forms 'successive instances' of presence (Lucy, 2004: 35). Whereas, Derrida's event 'exists

▶ **FIGURE 2.1**
The event in
metaphysics,
constructed from
Casati and Varzi
(2008)

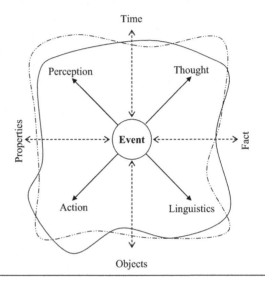

Event:-
- Important incident [the event-ness]
- Organised occasion
- Something that happens (philosophy)
- Single point in space and time (physics) '
(Encarta-Online-Dictionary)

——— Presence in space and time
—··—··—·· Event-ness, significance, important occurrence

outside any temporal order' (Kessel, 2007). Events are to come about, unexpectedly like a surprise, from the future. Kessel (2007) defines this future
as the 'impossible' future which is not the 'potential' future of this present.

> The event and the singularity of the event – this is what différance is all
> about. . . . So différance is a thought which wishes to yield to the immi-
> nence of what is coming or about to come: to the event.
>
> (Derrida, 1994: 252)

Simultaneously, the event which comes about sets off an interruption of 'con-
ventionally dominated' acts and contexts (Derrida, 1993). The event happens
unexpectedly and is 'absolutely different' from what is expected (Derrida,
1994). This 'différance' is produced through the difference between the event
presence/being in time and space and the other presence in time and space,
which helps to identify the singularity of the identity and meaning (Lucy, 2004).

> [The event] is another name for experience, which is always the experi-
> ence of the other . . . the experience of an event . . . is another name for
> the future itself.
>
> (Derrida, 1994: 252–253)

Finally, Derrida is particularly interested in the 'events that go unnoticed from within the metaphysics of presence', yet which represent life facts. However, they are not reducible to space-time presence/fact; they are always in action, fused with the four categories (Lucy, 2004: 35). Simultaneously, Kessel (2007) emphasises the differentiation between Derrida's event and the event which takes place in the 'present's state of affairs'.

Deconstructing binary opposition

In 'Of Grammatology', Derrida (1997b) considers the construction of western philosophy since Plato through binary oppositions (McQuillan, 2001). As part of deconstruction's continuous effort to deconstruct western metaphysics, it approaches binary oppositions in order to 'identify and undo'. This involves two steps: reversing the binary and displacing the binary oppositions so as not to 'involve binary logic at all' (McQuillan, 2001: 13). These steps are referred to by many authors as the stages of deconstruction in general; see (Alvesson and Skoldberg, 2000; Collins et al., 2005). However, it must be taken into consideration that the deconstruction of binary opposition does not involve the elimination of all binary oppositions. The deconstruction approach thus helps to define the limits within the binary opposition system to ensure that their representation of reality does not deceive (McQuillan, 2001). When I started writing about the role of deconstruction, I ran the risk of oversimplifying and/or overcomplicating the representation of deconstruction in this reading. This risk is increased in this section – the deconstruction of binary opposition. In an attempt, therefore, to develop a clear presentation, this part is presented through the categories of binary oppositions: definition, logocentrism, différance, and the supplement.

Binary opposition

Binary opposition divides 'conceptual material' into a pair of binary terms. These binary pairs are dependent on each other for meaning through 'difference', 'either/or' (Collins et al., 2005; McQuillan, 2001). However, these binary pairs are not opposites in reality; they are not equal; the first term is usually 'privileged', which is traditionally associated with masculinity (McQuillan, 2001). Through a stable differential relation between these pairs, binary oppositions are involved in philosophy and science as well as decision-making and thinking as they help to categorise 'objects, events, and relations' (Collins et al., 2005). However, McQuillan (2001) emphasises that binary oppositions are representations of 'western thought' rather than of reality, although they appear as part of this reality: a term is defined in opposition to what it is not, west/east, masculine/feminine, and so on. Thus deconstruction disturbs the logic of binary opposition (Collins et al., 2005). As discussed, this does not involve the elimination of binary oppositions but rather approaches the differential relation between the binary pair, i.e. the différance. Accordingly, 'only a displacement of a binary opposition can be said actually to undo that binary' (McQuillan, 2001: 18). To understand this, it is necessary to consider 'logocentrism' which constructs the binary logic of binary opposition. This leads to an exploration of 'différance' which considers the deconstruction of difference; the differential relation in between the binary pair.

Logocentrism

> Logocentrism constructs, or centres, sense and meaning, around the identity of these terms [binary oppositions] while disguising unresolvable tension between them.
>
> (McQuillan, 2001: 12)

Logocentrism is the logic of binary opposition; the inequality in between the binary pair and the privileging of the first term as 'positive'; the search for an ultimate and absolute truth, an origin (Collins et al., 2005; McQuillan, 2001). Accordingly, logocentrism considers the binary opposition pair; it privileges the first term and marginalises the second; and hence, considers the differential relation in between them as a process that starts from the first privileged term and moves towards the second (Collins et al., 2005: 46). The deconstruction of logocentrism is 'at pains to point out that it is impossible in principle to escape from logocentric thinking' (McQuillan, 2001: 13). We will discuss within 'différance' that both logocentrism and différance obscure each other as well as obscuring deconstruction.

Différance

> [Différance is] . . . the systematic play of differences, of traces of differences, of spacing by means of which elements are related to each other.
>
> (Derrida, 2004 [1979]: 27)

Différance deconstructs 'difference'. This raises two questions: what is 'difference'? How does 'différance' and 'deconstruction' approach this 'difference'? Difference sets up the logic of binary opposition in metaphysics. It considers the fixed differentiation between 'being this and being that' (Lucy, 2004: 7). The binary logic considers the dependency between the binary opposition pair through this fixed difference. For example, a binary term is either white or black. The identity of each, for instance the white, is dependent on its difference from the other, the black (Collins et al., 2005; McQuillan, 2001). Différance disturbs this notion of a fixed difference. It considers the production of difference between being this or that (Bennington and Derrida, 1993; Lucy, 2004). Accordingly, the deconstruction of binary opposition approaches the logic of binary opposition 'difference' through 'différance' rather than breaking up the terms of the binary (McQuillan, 2001).

> [F]or every element of the system only gets its identity in its difference from other elements, every element is in this way marked by all those it's not: it thus bears the trace of those other elements.
>
> (Bennington and Derrida, 1993: 74–75)

The definition of the French word 'différance' places it in between time and space and breaks down the pair. Différance is to 'defer' as 'an action of time', and to 'differ' as 'an action of space' (McQuillan, 2001: 17). This definition also recalls the event in which Derrida deconstructed its presence in space and time. The event does not exist; it is yet to come in the future, which is not the potential future of this present. And différance is 'the becoming-time of space and the becoming-space of time' (Derrida, 1982: 8). Différance thus involves a delay in time and space, a delay which implies 'that meaning is always antic-ipated or else re-established after the event' (Bennington and Derrida, 1993: 71–72). Accordingly, a binary term does not exist in the present; it extends between the 'past and future', which at the same time do not exist in the 'past and future' in reality (Bennington and Derrida, 1993). 'Différance is actively disruptive' (Collins et al., 2005: 77). You cannot pin it down, for the moment you do so, it is no longer différance but logocentrism (McQuillan, 2001). Différance, hence, produces binary difference. However, it is not fixed but instantly and continuously defers from this momentary 'presentation of difference' (McQuillan, 2001: 17). Différance is 'always in-between or in-the process-of' providing presence that is changeable and without 'being present itself' (Bennington and Derrida, 1993: 80; McQuillan, 2001: 17).

In summary, deconstruction does not break down the opposition between the binary terms but considers the différance of each term from the other as it appears to defer and differ (McQuillan, 2001). It deconstructs the 'possibility of any conceptual distinction' between the binary pair, 'can neither be a word nor a concept' (Bennington and Derrida, 1993: 70). This exclusion helps to define a binary term through a simultaneous definition of what a term is not; i.e. each term holds a trace of all the other terms which it is not (Bennington and Derrida, 1993; McQuillan, 2001). This displace-ment of the difference 'either/or' with double exclusion 'neither/nor' is rep-resented through Khōra in the next chapter. Khōra oscillates between two types of oscillation, double exclusion 'neither/nor' and participation 'both/ and'. This double oscillation helped to alienate Khōra from the binary oppo-sition logic.

Supplement

> [Deconstruction] is most commonly associated with post-structuralism but has its contextual roots in the historical conjunction surrounding structuralism. Like structuralism it is concerned with a certain idea of structure but it also, wishes to undo or desediment structures of all kinds, including structure of structuralism as well as structures older than structuralism.
>
> (McQuillan, 2001: 2)

As already discussed, binary oppositions help to organise and structure 'objects, events, and relations' as well as thought and decision-making. And différance is concerned with the production of the system of differ-ences between pairs of binary opposition. Différance initiates differential relations and simultaneously disturbs their stability. Thus, différance makes these structures, system of differences, possible; however, it also 'makes the idea of structure impossible' (McQuillan, 2001: 18). We shall now attempt to explore the 'structure', 'system', and the différance, deconstruction approach.

A systematic structure constitutes a fixed centre or origin 'referring to a point of presence' (Derrida, 2001 [1978]: 352; McQuillan, 2001). This idea of a structure built around a fixed centre is associated with Plato. This gave way to a 'de-centred' structure through structuralism (Lucy, 2004). Derrida differed from the structuralist de-centred approach which remained associated with the metaphysics of difference; centre/de-centre is another binary opposition (Lucy, 2004). The centre helps to close the structure; this closure allows systematic relations between defined elements. It limits and sets the boundaries of a structure which provides a clear definition of what is inside the structure and what is outside (McQuillan, 2001). Accordingly, the role of the centre is to manage the structure; it limits 'the play of its elements' inside the structure as well as 'orient[s], balance[s], and organise[s]' this structure (Derrida, 2001 [1978]: 352). However, as the centre closes up play within the structure, it 'opens up' and breaks outside the structure (Derrida, 2001 [1978]). This locates the 'traditional' centre both inside and outside the structure. The centre becomes the supplement which 'escapes the system and at the same time installs itself within the system to demonstrate the impossibility of the system' (McQuillan, 2001: 20).

Derrida continues to expose the determination of the centre in the history of 'structure', which through a system of difference developed 'a series of substitutions of centre for centre'. Accordingly, the presence of the centre is entitled to instant différance, to differ and defer, 'from itself into its own substitute' which questions the presence of the centre (Derrida, 2001 [1978]: 353, 354), and hence, the structure. Derrida deconstructed the 'centre' – the supplement – which is located both inside and outside the structure, breaking the boundaries and limits of the structure and thus deconstructing the structure in post-structuralism.

Meaning: triangle of signification

> [Traditionally] A sign . . . is . . . a sign-of a signifier referring to a signified, a signifier different from its signified, if one erases the radical difference between signifier and signified, it is the word 'signifier' . . . which must be abandoned as a metaphysical concept.
>
> (Derrida, 2001 [1978]: 355)

This section continues the discussion on the deconstruction of binary oppositions through différance, which Derrida inserted 'between signifier/signified, sensible/intelligible, word/concept' (Collins et al., 2005: 75). For Saussure et al. (1955), meaning is the product of the sign, as a 'permanent relationship' between the signifier and the signified as well as a play of difference between them (Collins et al., 2005: 62–63). The signifier is the physical image, sense, sensible, material, the body. And the signified is the mental image, thought, intelligible, concept, the soul. For Saussure, the meaning and identity of each sign is produced through a differential relation between the signs. This play of difference involves only the positive term, the signifier (McQuillan, 2001). Conversely, the relation between the signifier and the signified is stable; they cannot exist without each other (Collins et al.,

2005; McQuillan, 2001). However, this relation is arbitrary. Accordingly, 'the concept is fixed as the signified and has priority over its arbitrary and conventional mode of expression as a signifier' (McQuillan, 2001: 18). For Derrida, on the other hand, as discussed in the previous section, différance does not consider the difference between the signs but the system of differences. The sign refers to a concept which itself refers to reality (Bennington and Derrida, 1993). The sign refers to an absent signifier, which refers to another signified through différance (McQuillan, 2001); i.e. the signifier/signified relation is no longer stable, and the signified is caught in differential relations to other concepts, through its signifier expression.

> [Différance] . . . holds that an element functions and signifies, takes or conveys meaning, only by referring to another past or future element in an economy of traces.
>
> (Derrida, 2004 [1979]: 29)

Embracing the margin

> The very condition of a deconstruction may be at work, within the system to be deconstructed; it may be already located there, already at work, not at the centre but in an excentre, in a corner.
>
> (Derrida and De Man, 1989: 73)

McQuillan introduced 'embracing the margin' as a 'deconstructive method', if such a thing exists, as previously discussed in the 'pas de méthode' section. The deconstruction of a discourse lies in the margin of this discourse. The margin is subservient, dominated, excluded by the dominant and more important binary at the centre of the discourse. Accordingly, this domination of the margin opens up the discourse for deconstruction; the 'position of the margin . . . is responsible for the entire structure' (McQuillan, 2001: 30). However, in continuation of discussion on the différance, we should be careful not to reverse the binary; that is not to include the dominated and exclude the dominant, which would continue the binary logic in different terms.

> As a cornerstone, it supports it [the structure of the discourse], however rickety it may be, and brings together at a single point all its forces and tensions. It does not do this from a central commanding point, like a keystone; but it also does it laterally, in its corner.
>
> (Derrida and De Man, 1989: 73–74)

The cornerstone represents a set of binary oppositions which is marginalised from the discourse. Figure 2.2 represents a visual metaphor of Derrida's reading of the cornerstone in McQuillan (2001), which is outside the centre of the discourse, the arch. The cornerstone is like the 'keystone' which holds the structure of the arch and hence, responsible for its stability. A defective keystone entails the failure of the arch structure. However, the cornerstone is ex-centric. It both holds the arch structure from its location at the corner,

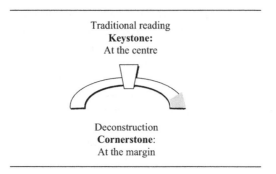

▶ **FIGURE 2.2**
Deconstruction is
interested in the
cornerstone at
the margin of the
discourse rather
than the keystone at
the centre

and is simultaneously defective and responsible for the arch collapse. This
interest in the margin reflects deconstruction's interest in the instability of
the whole discourse. The margin attributed to both the construction and
deconstruction of the discourse is readily described as 'defective cornerstone'. The cornerstone stands for the whole structure, while pushed to the
margin. It is considered the first stone, upon which the whole structure is
built. However, it is defective and that makes it unstable. The reader's task is
to find the cornerstone which will readily deconstruct the whole structure
(McQuillan, 2001).

> Deconstruction is interested in this so-called marginal term. Giving that
> this exclusion does not reflect lived reality but is an operation of power
> enacted on behalf of a certain politically constituted group then the recu-
> peration of the margins is a necessary step in demonstrating the injus-
> tices which are disguised by the work of logocentrism.
>
> (McQuillan, 2001: 23)

McQuillan (2001: 28) demonstrated reading strategies to embrace the
margin and deconstruct the text through a 'singular path of a moment
reading'; a 'specific and situational' reading which is approached from
inside the discourse, rather than applying an external methodology. Consequently, he emphasised that this is not a subjective reading that puts
meaning into the discourse through 'picking on an obscure detail'. It is a
reading of the text, which has already privileged the discourse in the centre
and excluded the other in the margin; a reading which sets a play of différance inside the text, against its inherent structure. However, this is an
essential step towards deconstruction that is continued through a personal
reading; a reading which puts the reader inside the text, for the singular
moment of reading and leaves a trace of the reader in the text. Accordingly, we recognise the uniqueness of the event of reading, which differs
every time the text is read. Accordingly, this is followed by a 'movement
from specific to general . . . along a singular path of reading' (McQuillan,
2001: 26). This reading considers the wider context of the text; the production and functioning of the text binary oppositions within their culture; and demonstrates the instability of their production and hence puts
the cultural context at risk. The reader is hence enabled to deconstruct

'the network of power relations' that is based on and simultaneously constructed from these oppositions.

A review of deconstruction strategies was presented here; deconstruction is simultaneously interested in architecture as a way of thinking, which will be explored further in the next section. This involves the reading of architecture presence, boundaries, and structure as traditionally defined by Vitruvius tradition, the displacements of these boundaries, as well as the reflections on its structure, inside/outside discourse.

ARCHÉ-SPACE: VITRUVIUS VS. DERRIDA[3]

> Vitruvius was the first . . . to cover the entire field of architecture in systematic form.
>
> (Kruft, 1994: 21)

In architecture, Pollio Vitruvius (70/80–15 BC) is a key figure, and one who has been considered the 'most influential' writer (Markus and Cameron, 2002: 19). Architecture classicism 'begins with Vitruvius' (Mallgrave, 2006: 8), to whose text *Vitruvius: The Ten Books on Architecture* (Vitruvius, 1960), 'at least until the eighteenth century, all other texts referred' (McEwen, 2003: 1). And architecture space was thus strongly related to and represented through his famous triad Firmitas (structure), Utilitas (function), and Venustas (aesthetics). Many contemporary architecture writers still believe in the power of Vitruvius's triad, for example Markus and Cameron (2002). However, the dominance of this triad, which is shaped within the context of the Roman city, is questioned by architecture historians, 'what is a modern historian to make of architecture so circumscribed?' (McEwen, 2003: 301). Simultaneously, Tschumi (1980) traced the development of architecture beyond this circumscription, as he observed it and questioned the unrecognised changing limits within architecture space, through space, movement, and event (see Table 2.1): 'does architecture fail to realise the displacement of limits it has held for long?' (Tschumi, 1981a: 108).

> For Derrida, [architecture deconstruction] means interrogating architecture's 'traditional sanctions' that buildings should be useful [Utilitas], beautiful [Venustas] and inhabitable [Firmitas].
>
> (Collins et al., 2005: 131)

McEwen (2003: 301) reflects on the current 'trend . . . to bypass master narratives and focus instead on the fragmentary, the subversive, the marginal, the feminine'. This trend is strongly associated with post-structuralism and particularly with Derrida's deconstruction project. However, evidently Derrida did not bypass Vitruvius and other master narratives on architecture. On the contrary he questioned these traditionally established 'sanctions' in architecture theory and literature (Collins et al., 2005). He also considered the destabilisation of the authorship, title, edges, and boundaries in architecture through the deconstruction of these 'master narratives'.

▶ **TABLE 2.1**
Arché-space:
Vitruvius vs.
Derrida

Architecture presence	Vitruvius 'Firmitas, Utilitas, Venustas'		
Boundaries: regulations and norms	**'Firmitas,**	**Utilitas,**	**Venustas'**
	Structure stability and building material	Appropriate spatial accommodation	Aesthetics, harmony, symmetry, and unity
1960s	Anything could be built provided that you could pay for it	**Function** (user activity programme)	Structural linguistics **Form** (style and geometry)
Markus	Formal rules **space** (place morphology and organisation)		

'The twentieth-century has disrupted the Vitruvian trilogy' (Tschumi, 1981: 108).

Tschumi (minimalist)	The accident is the norm and continuity the exception	'Body-in-**space**', 'space-of-**movement**', 'intrusion of **events** into architecture spaces'	**Destabilisation**: today we favour a sensibility of the disappearance of unstable images
(maximalist)		**Concept** (space), **Content** (movement), **Context** (event)	

Architecture Deconstruction	Boundaries and limits
	Structure (inside/outside)
	Derrida

~~Firmitas~~ ~~Venustas~~ ~~Utilitas~~ Space movement event

> Vitruvius's authoritative voice from the past . . . defined for all time what the important issues in architecture were.
>
> (McEwen, 2003: 2)

This section approaches two readings of Vitruvius's trilogy by Thomas Markus and Bernard Tschumi (Figure 2.3). Markus (1987: 19) on the one hand, emphasises the role of Vitruvius in architecture today: 'Vitruvius is alive and well' and his famous trilogy known as *Firmness, Commodity, and Delight* is still present in today's architectural theoretical discussions. It constitutes 'themes as language and human body' (Markus and Cameron, 2002: 19, 21). Consequently, Markus echoes Vitruvius's trilogy in his

Vitruvius		
Firmitas Structure Firmness	Utilitas Function Commodity	Venustas Aesthetics Delight
Thomas Markus		
Construction	Utility	Beauty
Bernard Tschumi		
Structure Stability	Appropriate spatial accommodation	Attractive appearance

◀ **FIGURE 2.3**
Vitruvius triad,
readings and
interpretations

triad of space, form, and function. Tschumi, on the other hand, questions the dominance of the Vitruvian triad on contemporary architecture theory and negotiates its diminution throughout history: 'The twentieth century has disrupted the Vitruvius trilogy' (Tschumi, 1981a: 108). Accordingly, he explores the development of architecture discourse from Vitruvius's trilogy towards a 'combinations of spaces, events, and movements' (Tschumi, 2001: 255). Simultaneously, he displaces this 'minimalist' triad through another maximalist one that involves 'concept, context and content' (Tschumi, 2004; Tschumi, 2010). This section therefore studies the development, or more accurately the displacement, of these boundaries through the reading of Tschumi.

Firmitas

Vitruvius's 'Firmitas' considered the stability of the structure and foundation as well as the appropriate use of building materials (Mallgrave, 2006: 8–9). This consensus was displaced by another in the 1960s which declared 'that anything could be built provided that you could pay for it' (Tschumi, 1980: 109). Hence, by the 1980s the concern for structural stability was displaced by formal considerations. The latter displacement was echoed in Markus's 'space', which considers the morphology, typology, and organisation as the formal properties of space (Markus, 1982b; Markus, 1987). Today, Firmitas, 'solidity, firmness, structure, and hierarchy', is displaced again by relativity, quantum theory, and uncertainty; the new displacement hence calls for a 'new regulation', where 'the accident [is] the norm . . . and continuity the exception' (Tschumi, 1989: 218–219).

Venustas

Simultaneously, 'Venustas' considered aesthetic rules through the lens of symmetry, harmony, unity, order, and arrangement inspired by the perfection of the human body, 'all in line with Pythagorean and Platonic principles' (Markus and Cameron, 2002: 19). These strictly physical considerations of aesthetics were displaced by structuralist and language preferences, which introduced architecture of metaphors and symbols. This approach is reflected in Markus's triad through the architecture form, defined as style and composition (1982b; Markus, 1987). On the one hand, style involves 'symbolic, semiotic and abstract content of style' (Markus, 1982a: 5), and

on the other, architectural composition involves formal 'geometric properties, the proportions, articulation, colour, ornamentation, and surface
treatment' (Markus, 1987: 468). Again, Tschumi (1989: 217) highlights the
displacement of these physical and linguistic structural 'preferences' of a
static, balanced, symmetrical, and harmonious architectural image towards
destabilisation. 'Preferences are changed not as a style but as a destabilization'. These new preferences are hence, traced in architecture 'disjunction,
dislocations, deconstruction'.

Utilitas

The use and definition of this term Utilitas is highly debatable. For Vitruvius, 'Utilitas' represented the 'appropriation of spatial accommodation'
(Tschumi, 1980: 108), the appropriate arrangement of spaces to accommodate and facilitate the use-of-space, which includes activities, as well as
economic and construction considerations (Mallgrave, 2006). Accordingly,
'Utilitas', in reality, refers to the space-of-Utilitas, rather than the Utilitas-in-
space. Markus (1982a, 1986, 1987) adopts a similar definition of the space-of-
'function'. Again, form and space accommodate the 'function experience', as
defined through the 'functional statement' in the architectural programme
and which has been simultaneously defined by society (Markus, 1982a: 5;
Markus, 1986: 486). This entails a cause-and-effect relation between function and space, body and space, people and place. However, Tschumi (2001:
4) argues that 'in contemporary urban society, any cause-and-effect relationship between form, use, function, and socioeconomic structure has
become both impossible and obsolete'. He also debates the complicated
displacement 'from a space-of-body to the body-in-space' (Tschumi, 1981a:
110–111). This displacement involves both the movement of the body in
space and the generation of 'spaces of movement'. These generated spaces
help the 'articulation between the space of senses [body/matter] and the
space of society [social context]'. Simultaneously, this articulation entails
the 'intrusion of events into architecture spaces', which are 'independent but
inseparable from the spaces that enclose them' (Tschumi, 1981a: 111). Consequently, Tschumi distinguishes the event from function, either the function in space or the space of function.

In-context: minimalist vs. maximalist

> To bring context and content to event and movement is a way to con-
> front them with the realities of both culture and production.
>
> (Tschumi in Michele, 2009: 29)

Both Vitruvius's and Markus's trilogies considered a rational approach
to defining the boundaries and limits of architecture, norms, regulations, and guidelines which were to apply to architecture. Simultaneously,
Tschumi's triad highlighted the displacement of these triads in twentieth-
century architecture. He also, emphasised the obsoleteness of the rational
cause-and-effect relationship between the three components of both triads. Accordingly, his triad space, movement, and event brought about the

destabilisation of the architectural structure and aesthetic image, and the displacement of the space of function (for more details, refer to Tschumi, 1994b *Manhattan Transcripts*). It is necessary to highlight the duality of this displacement, the body-in-space and the space-of movement, which helps both to blur the distinction between the components in both Vitruvius's and Markus's trilogy and to bring about the 'event' between them.

Subsequently, Tschumi reflects on the schematic displacement of the 'old trilogy' through 'mental, physical, and social space or, alternatively . . . language, matter and body', which simultaneously responds to the 'conceived, perceived, experienced' architecture space (Figure 2.4). These categories recall 'space/place' in philosophy – namely mind, matter, and body. However, he also questions the authority of these categories or components as limits and norms, 'but can distort them at will' (Tschumi, 1981a: 111, 112).

Tschumi (1980) reflects on yet another phenomenon related to twentieth-century architecture, namely the reduction of architecture space into two streams, 'maximalist' and 'minimalist'. The latter concentrates on the details in architecture, style, technique, and so on. The maximalist, on the other hand, extends to the urban context, social, cultural, political, as well as the programme. Accordingly, he negates the separation between these two streams, which are evident in modern and postmodern architecture. Consequently, he complements the 'new trilogy' and his own trilogy, 'Space, Event, Movement', with another trilogy of 'concept, context, and content' (Tschumi, 2004) (Figure 2.5). Tschumi (2004) emphasises that there is no architecture without concept, content, or context (Michele, 2009, Walker, 2006). Hence, he introduces the concept-form which precedes content and context; architectural space is 'a neutral container [that] can house any number of activities' (Tschumi, 2004: 12; Tschumi, 2010) Through the concept-form, he rejects modern functionalism; form does not follow function, and

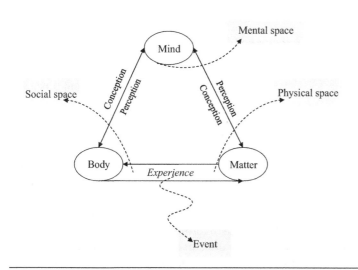

◀ **FIGURE 2.4**
Vitruvius's trilogy displaced and distorted as perceived by Tschumi

▶ **FIGURE 2.5**
Tschumi's
complementary
trilogies: Space,
Movement, Event
vs. concept,
content, context

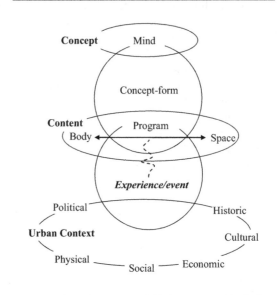

concept does not follow content. Accordingly, the concept-form brings a high level of abstraction to architecture space: 'complexity that includes materials, movement and programmes' (Tschumi in Michele, 2009: 12). At the same time, he differentiates between his approach to context and 'contextualism', which implied 'aesthetic conservatism . . . in the 1980s and 1990s'. For Tschumi, the context exceeds the physical and visual setting to include the 'historical, geographical, cultural, political or economic' urban context (Tschumi, 2004: 11). Finally, Tschumi (1994b: xx) argues against any static definition of architecture space, boundaries and structure, 'Architecture: a form of knowledge whose limits are constantly questioned', as I shall discuss in the following section.

Architecture presence/deconstruction

> [T]he body . . . the starting point and point of arrival of architecture.
> (Tschumi, 1980: 110)

'Vitruvius's project has been a normative one, motivated by the desire for rational systematization' (McEwen, 2003: 3). Architecture thus revolves around a fixed theoretical core and the boundaries are well defined. Significantly, the body has been present at the centre of the architecture discourse. It thus limits and sets the boundaries of a well-defined structure, the 'rational systematization' of the discourse. However, as discussed in the previous chapter, this presence has in reality subverted the body through both modern functionalism and postmodern formalism. Accordingly, the body is the supplement of the architecture discourse that escapes outside it. It is thus caught in a differential relation, both inside and outside and simultaneously neither inside nor outside architecture space/discourse. Furthermore, the displacement and alteration of Vitruvius's boundaries have been simultaneously reviewed through Tschumi's trilogies of space, movement, and event, and concept, content, and context. Tschumi refuses a static definition of architecture space and boundaries, and considers their displacement as a continuous process. Architecture no longer revolves around one static core but around multiple cores 'around which issues revolve and occasionally intersect: space, programme, body, envelopes, global versus local, economy of means, typology versus topology, concept form, and so on' (Michele, 2009: 27).

I have also reflected on architecture space as a space of knowledge of physical space in context in the preliminary reading, and accordingly explored the development of architecture space in close relation to philosophy. This review examined the tradition which regards philosophy as an outside and simultaneously transcendent space of knowledge to the practice of architecture, art, and science. Controversially, Mullarkey (2006) challenged philosophy to accept 'new philosophical thoughts' through these practices. This challenge is brought about and made possible through his reading of transcendence in relation to immanence.

On the one hand, the transcendent knowledge 'is outside both literally and figuratively' as well as 'multiple and relative'. At the same time, the immanence of the practice, or physical space, in the space of knowledge is in a continuous process of change; '[the philosophical] frame, the place where one takes a stand, is never permanent' (Mullarkey, 2006: 193). Accordingly, the transcendent knowledge, roaming outside, is only approachable from the inside through a temporal frame of immanence. This reading breaks the boundaries between inside and outside knowledge/practice, which have previously elevated philosophy and theory to an outside space of knowledge to transcend practice. The dependence of the transcendent on the immanent makes it appear to be inside the boundaries; knowledge transcends from the inside the practice, the physical space. Simultaneously, this reading could be projected on to architecture/urban space: boundaries and limits, inside/outside, content/context.

ARCHÉ-DECONSTRUCTION READING STRATEGIES

> [D]econstructionist architecture is there; there is a lot of it about and there is more to come.
>
> (Broadbent, 1991a: 11)

We have reviewed deconstruction strategies through McQuillan's (2001) representation of 'five strategies of deconstruction', pas de méthode, deconstruction in context, a history without history, deconstructing binary opposition, and embracing the margin, and induced the two other sub-strategies, event and meaning. Also, this involved repeated discussion of the difficulty of approaching deconstruction without the risk of oversimplifying and/or simultaneously overcomplicating its representation. The reality is that a life-time could be spent studying just one strategy associated with deconstruction. Accordingly, an ellipsis of this reading of these strategies is developed. It is necessary to emphasise that this is an ellipsis of deconstruction strategies presented in this book rather than an ellipsis of deconstruction.

First, deconstruction is not a method, or a set of rules, applied to interpretation (Lucy, 2004). These reading strategies essentially work from inside the text content. Second, deconstruction is totally dependent on the nature of this content. It is also dependent on the author/reader as they acknowledge the multiplicities of meaning and truth. Third, this dependence on the content represents a trace of contextual reality and history of both content and context. The fourth strategy, deconstruction of binary oppositions, is referred to by many authors as the stages of deconstruction in general (see Alvesson and Skoldberg, 2000; Collins et al., 2005). Binary opposition divides 'conceptual material' into a pair of binary terms, e.g. west/east, masculine/feminine. These binary pairs are dependent on each other for meaning through 'difference': a fixed 'either/or' relation (Alvesson and Skoldberg, 2000; Collins et al., 2005). However, these binary pairs are not opposites in reality; they are not equal; the first term is usually 'privileged', which is traditionally associated with masculinity (McQuillan, 2001). The deconstruction of this binary thus involves two steps: reversing the binary and displacing the binary oppositions so as not to 'involve binary logic at all' through 'this thing called différance' (McQuillan, 2001: 13, 19). Différance displaces the stable either/or relation with the dynamic oscillation between neither/nor and both/and relations. Accordingly, it deconstructs this hierarchy through the simultaneous presence of two conflicting ideas, thus destabilising the notion of 'fixed' meaning and creating potential for alternative interpretations (Alvesson and Skoldberg, 2000). The fifth strategy embraces the margin, which denies the proposed representation and brings the marginal to be the centre. Deconstruction is thus interested in the marginalised set of binary oppositions, which is called the cornerstone. The cornerstone is considered the first stone, which is not at the centre and upon which the whole structure of a representation is built. However, it is defective and that makes it unstable. The reader's task is thus to find the cornerstone which will readily deconstruct the structure of the proposed representation (McQuillan, 2001). At the same time, the subversion of the hierarchy between centre and margin blurs the peripheries of the content. The centre escapes the boundaries and is thus located both inside and outside the content, which further destabilises the established representations (Derrida, 2001 [1978]; McQuillan, 2001).

Reading singularity: between meaning and event

Sloterdijk (2009: x) demonstrates that there are 'essentially only two' readings: 'singular' which is more concerned with the content, and 'in context' which extends to include the context. On the one hand, a singular reading is primarily concerned with the content, the internal structure, composition, style, and details as well as emphasising the personal space of the reader/

user. However, it risks the marginalisation of the context. A contextual reading, on the other hand, places the content in relation to its (urban) context: historical, geographical, cultural, political, and/or economic, in order to understand the meaning allocated in the context. This reading thus emphasises the socio-cultural space through typologies and symbolic relations. However, it risks the marginalisation of content, being both 'too easy to deal with' and subverted by the context (Sloterdijk, 2009: xii). As previously discussed, architecture/urban space is simultaneously perceived as minimalist and maximalist, (Tschumi, 1980); and social space as personal and socio-cultural, respectively. Accordingly, I shall consider these readings under the headings 'singular' and 'contextual'.

Deconstruction strategies are intrinsically dependent on the author/ reader as they acknowledge the multiplicities of meaning and truth. This in turn is consecutively dependent on the singularity of each and every reading. However, it should be noted that Derrida's singular reading is not perceived as in opposition to the contextual. In a singular reading, he recognises the inscription of this context inside the content, while simultaneously rejecting the transcendence of the context to the content. The boundaries inside/outside the content of the discourse are blurred through the concepts of the 'trace', and the 'différance'. The unity of meaning is thus displaced by the singularity of reading, and hence these strategies are dependent on the reading context, which is never the same.

Accordingly, it could be said that reading the discourse through these deconstruction strategies is a reading event which is not pre-planned and is therefore unexpected and yet to happen through a singular reading context, and the meaning is yet to be established after the event. However, meaning does not displace the event nor is displaced by it. Controversially, it appears that the reading event produces the discourse meaning. This sentence should be taken cautiously, as it could imply a cause-and-effect relation between event and meaning, which would disturb the understanding of both. In summary, Derrida's deconstruction favours the 'singular reading' through time and space rather than the contextual, a reading which is influenced by the discourse, content, context, and author (McQuillan, 2001). Consequently, this reading singularity helps the development of two simultaneous strategy triads through meaning and event. Importantly these should not be taken as strategies for application but as approaches to help the critical interpretation of the discourse. Meaning, in relation to strategies involving the deconstruction of binary oppositions and the cornerstone, emphasises the deconstruction of the boundaries of the urban space discourse. And the event, in relation to a-history and context reading strategies, emphasises the deconstruction of boundaries of time, i.e. the reading singularity of the discourse approaches the deconstruction of the discourse boundaries through différance in time and space.

Meaning: binary oppositions: cornerstone

Meaning is traditionally fixed through the metaphysics of difference. Both meaning and cornerstone are the by-product of the binary logic; signifier/

signified, centre/margin, keystone/cornerstone. Binary logic considers the inequality between the two binaries privileging the first term. Meaning is produced through the fixed difference between the signs/the signifiers; an 'either/or' relationship. Deconstruction is not interested in the destruction of these binaries; deconstruction considers the limitations of their representation through a fixed singular meaning at the centre of the discourse. Deconstruction 'makes every identity [meaning] at once itself and different of itself' (Royle, 2000: 11). Meaning is no longer fixed; it is continuously changing through différance. Différance helps the displacement of the fixed 'either/or' relation through the oscillation between neither/nor and both/and, which obscures determinacy, fixed meaning, and relations. Meaning is thus caught in a continuous differential relation between the signified, which traces other signs. It is displaced through a system of différance between the signs, giving way to a reading singularity.

Subsequently, deconstruction is already at work inside the discourse through the 'defective cornerstone'. It acts as 'a theoretical and practical parasitism or virology' (Royle, 2000: 11). However, the cornerstone works at the margin of the discourse and not at the centre. A deconstruction reading would reach out for the cornerstone which differs from one singular reading to another. Consequently, the deconstruction of the inside/outside binary through the displacement of the centres of the discourse that escapes and blurs the established structure and boundaries of the discourse, and meaning is thus caught through the reading singularity of the cornerstone.

Event: context: history

The event is intrinsically an instance of presence in space and time, which is yet to come in the future, isolated from other successive events (Derrida, 1994). Simultaneously, the event is both traced by and a trace of the discourse content, context, and author, the potential reader. Accordingly, the reading event both traces and is traced by what is happening outside the discourse: the context, society, politics, economics, history, and so on. At the same time, the event rejects the transcendence of the context and author outside the discourse; both are inscribed inside the discourse breaking the boundaries between inside/outside, as well as between each other.

Consequently, deconstruction rejects the idea of the 'history' which constitutes both the production and explanation of the discourse (McQuillan, 2001). The reading event is contextual although it recognises the impossibility of building a past context; the event is yet to come in a future that traces both the present and the past without following their traditional path (Derrida, 1994). Hence, it emphasises the blurring process between the boundaries of time, past, future, and present.

Lastly, I refer back to Royle's comprehensive definition of deconstruction, if such a thing exists, where an expected comment from Derrida might be that, 'deconstruction is all that and yet none of those'. Deconstruction is:

> Not what you think: the experience of the impossible: what remains to be thought: a logic of destabilization always already on the move in

'things themselves': what makes every identity at once itself and different of itself: a logic of spectrality: a theoretical and practical parasitism or virology: what is happening today in what is called society, politics, diplomacy, economics, historical reality, and so on: the opening of the future itself.

(Royle, 2000: 11)

NOTES

1 The title of this chapter is taken after Derrida's (in Wolfreys, 2007: 72) definition of the 'arché' in his article 'Khôra' '(That from which one starts, the beginning, the start, the chief, the principal or first in authority), as generally in architecture, architectonics, arche-écriture or arche-writing'.
2 Supplement is the deconstruction of the inside/outside binary opposition. This will be discussed later in this chapter.
3 I represented an earlier version of 'Vitruvius to Derrida: towards re-reading architecture space', at the fourth EAAE-ENHSA sub-network workshop on architecture theory, 'Architecture Theory: A technical practice', Fribourg Switzerland, 14–16 October 2009.

Chapter 3

CAIRO-KHŌRA

Here is my idea: design chóra, the impossible place: design it.
(Derrida in Casey, 1997a: 312)

Derrida's 'awkward interaction' with Tschumi and Eisenman in 'Parc de La Villette' brought about a controversial dialogue between architecture and deconstruction. Accordingly, Derrida proposed 'chóra', to build chóra, as presented in his, by that time incomplete, essay on Plato's Chóra, as the design theme (Morgan, 2006). However, 'chóra is precisely what cannot be designed' (Casey, 1997a: 312). Derrida simultaneously engaged with another 'awkward interaction' upon his visit to Cairo, where he reflected on the misrepresentation of the city urban space through the mono-lithic dominant reading of the Islamic city. As discussed, this misrepresentation is manifested through the dynamics of the city's public space and has led to a growing tension in between Cairenes and their urban space. This helped to raise question(s) about the identity of Cairo – who is Cairo?

These interactions have simultaneously raised many questions and/or debates about the reading of architecture and urban space. The potentials for addressing these controversies can be traced within Derrida's descrip-tion of deconstruction as 'the opening of the future itself, a future which does not allow itself to be modalised or modified into the form of the pres-ent' (Derrida, 1992: 200). In this sense, deconstruction could help to develop a new reading of urban space in general and of Cairo in particular, which does not repeat a present or past reading and representation.

I shall thus approach a deconstruction reading of urban space, employ-ing a virtual walk through Cairo temporal, formal, public space. Simulta-neously, I shall explore Derrida's deconstruction reading of space, chóra: 'Khōra'. I follow his question: who is Khōra? and accordingly represented her in this book as a third genre, promoted uncertainty through oscillation, connoted her to architecture as an impossible space, and accordingly a par-adigm shift in the understanding of space/place. Consequently, I argue that Khōra inhabits the urban space as urban space inhabits Khōra; they are both the abstraction and realisation of each other through space/place, without building that space/place.

WHO IS CAIRO?

> One day I would like to write a book on Egypt and Derrida. The two . . .
> are tethered together in suggestive ways.
>
> (McQuillan, 2010: 255)

This preliminary reading is developed through my personal experience in 'Cairo, my city'. In addition, I extract a primary reading of Cairo space from newspapers, internet blogs, and ethnographic reports of the city, as well as from unstructured conversations with various government officials. The main aim of these conversations/interviews is to explore and attempt to develop a general idea about Cairo space; consequently, the main question is: 'what is happening from your perspective?' In this reading, I attempt to identify the binaries inherited in the city and the display of power beyond these representations. The aim here is to highlight the cornerstone that helped to associate the city to a monolithic 'oriental' image and identity. This helps to open the way to a reading deconstruction event that appreciates the dynamic realities and complexities in the city urban space. Accordingly, this reading draws on a virtual walk through the city that comprises three parts. The first is a virtual walk in time since the city's foundation in AD 969 continued through the Islamic, the modern, the presidential. The second is a spatial walk through the city's public space, where the different phases in time are reflected in this space. Third, a formal walk investigates the institutional organisations and how these are reflected in the public space.

Temporal space

> The history of Cairo and its discontinuities have thus brought together
> urban areas that differ widely in their concepts, their economic role,
> and the social and cultural level of their residents. Such contrasts . . .
> are also present in all great modern metropolises, but they are partic-
> ularly distinct here.
>
> (Raymond, 2007 p. 364–365)

This section reviews the city development in time since its foundation by the Fatimid. However, it is necessary to highlight the fact that Cairo's history is built on the old Egyptian heritage of Pharaohs, and on the Roman, the Coptic, and the Byzantine Heritage. Since its foundation three main periods can be identified, which define the city's characteristics and evolution: the Islamic city, the modern city, and the current presidential city. These three periods were disrupted by two main events, which stopped the continuity of the development of the city. These were the French expedition of 1798 and the military revolution of 1952. Another event is the revolution on 25 January 2011. However, this event has caused a disturbance in the presidential city rather than disrupt its continuity (see Figure 3.1).

Originally Cairo was called Al-Mansurriya after the North African Fatimid capital. It was Al-Muez Ledin Allah Al-Fatimy who called the city, Cairo, and this became well known as Fatimid Cairo. Today this part of

▶ **FIGURE 3.1**
Continuities and
discontinuities in
the city history

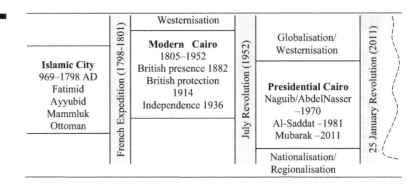

the city is known as the historic core or sometimes as the Islamic city. This city was developed through four dynasties: the Fatimid, Ayyubid, Mamluk, and Ottoman 969~1798. The continuity of the history of the 'Islamic' city was hence interrupted by the French Expedition of 1798–1801, which led to the development of modern Cairo by the Mohamed Ali family from 1805 to 1952. It could be said that the history of the development of modern Cairo was chiefly influenced by two main characters. Mohamed Ali Pasha (1805) established the modern city and introduced new institutional, social, and economic systems. Ismail Pasha, 1863–1879, established a comprehensive plan for the redevelopment of the city which projected the city's 'dream of westernization'; 'street has primacy, grid urban geometry . . . as well as new western architecture model' (Raymond, 2007: 309). This projection of a western image on the city continued throughout the British presence in 1882 and protection in 1914 until the city's declaration of independence in 1936. Through this period Raymond (2007) calls Cairo the 'British city'.

The continuity of city history was altered again by the 1952 revolution and the appointment of the first Egyptian president Mohamed Naguib, soon followed by Gamal Abdel Nasser, 1956–1970, Mohamed Anwar El-Sadat, 1970–1981, and Mohamed Hosney Mubarak, 1981–2011. It is hard to give a short account of this long and rich period of Egypt's history. Accordingly, the main ideologies that helped the city development throughout this time are analysed and presented here. Abdel Nasser's 'nationalist' approach and socialist system (Adham, 2004) pronounced the western model as unsuitable for the city of Cairo (Salheen, 2001), and the city searched for a traditionalist and/or regionalist model to replace it. Then the post-war policies of the El-Sadat presidency promoted an open market economy and the western model came back into favour (Salheen, 2001). The city suffered a massive increase in population density (Singerman and Amar, 2006a) and experienced a widening gap between the higher social class residential areas and the deteriorating lower-income areas (Salheen, 2001). Both cities – the nationalised and westernised/globalised – and their problems continued to grow through the presidency of Mubarak, and the dreams of nationalism and westernisation continued to co-exist within Cairo space. The city grew into a massive metropolis comprising both historic Cairo and modern

Cairo. Recently, the return of Cairene immigrants from the western and the Gulf countries has encouraged the emergence of New Cairo, a place that includes new (gated) communities which reflect a different/contemporary lifestyle from that of the city of Cairo itself. Overall, the layers of the city history developed over time, not replacing each other but rather changing and developing to co-exist in the contemporary city. However, the continuity of this presidential city was interrupted again by the 25 January revolution in 2011. Significantly, the revolution aimed to build the future city identity rather than restore the lost identity, which was the aim of the July 1952 revolution. This helped to re-emphasise the question of the place identity – who is Cairo?

> But as events in Tahrir square continue to unfold it becomes ever clearer that such social processes are intimately grounded within space – and that specific places within the city take on highly symbolic and ideological meanings. This confirms that urban space is central to the construction and re-construction of civic and national identities.
>
> (Kellett and Hérnandez-Garcia, 2013: 14)

Public space

Interestingly, these three periods were projected onto the city's planning and development of public space and green areas. Rabbat (2004) presents a review of urban space development in the three periods, and demonstrates the contradiction between the common image of a city developed in the desert, and the fact that Fatimid Cairo was founded around the 'Bustan Kafur'. A 'bustan' is the traditional equivalent of western parks; however, a 'bustan' is essentially productive with fruits and vegetables. Today, the location of 'Bustan Kafur' is occupied by Al-Musky, one of the most crowded neighbourhoods in Cairo. Another type of urban space of this period is the 'mydan', the equivalent of today's square. However, these squares were not accessible to the public. Rabbat (2004) also reflected on the 'conceptual distinction between private and public space' that was functionally projected; private spaces involved leisure and entertainment and public spaces involved commerce and worship. Furthermore, the city continued to produce private gardens inside the walls of palaces, which helped to create the image of the 'overbuilt city' in the desert with few open spaces.

Modern Cairo, the second historical period of the city, witnessed the development of most of her gardens, squares, and parks especially in the time of Ismail Pasha, for example Azbakiya garden and Horreya garden (El-Messiri, 2004; Rabbat, 2004). These parks and squares projected a western image, and were built for the royal family, westerners, and elite Egyptians who could afford the entrance fees (El-Messiri, 2004). This helped to attach the image of the westernised city to an elitist and well-developed identity. Although many of these places still exist with the city urban fabric today, many were taken over by urban developments, especially after the state bankruptcy caused by the extravagant spending of both Said (1854–1863) and Ismail (1863–1879) (Rabbat, 2004).

Following the 1952 revolution, these parks and squares were opened to the public for free. However, 'attention was directed to the internal affairs'; and the development of the city infrastructure, transportation, and housing was given priority over gardens and parks (El-Messiri, 2004: 226; Rashed, 2005). The 'Obelisk Garden', built during Abdel Nasser's presidency, is considered the only significant park built up until the 1980s (Rabbat, 2004). Also, the 'open economy policy' promoted the privatisation of the remaining palatial gardens and the Nile promenade, and their ownership was transferred to expensive hotels, clubs, and restaurants (Rabbat, 2004).

The growing population and urban development continued to eat up the remaining green areas and public spaces piece by piece. By the 1980s, awareness was growing of the consequences of these developments on the environment. Accordingly, the Cairo government established new plans to address the scarcity of green areas and public spaces in order to enhance the quality of urban life within the city (El-Messiri, 2004). These projects were initiated in 1987, with the establishment of 'The International Garden' in Madinet Nasr through donations from several countries (El-Messiri, 2004) (Figure 3.2).

Each of these countries chose a location to bear its name in the garden and was responsible for the design and making of this area. Two new parks, 'Al-Fustat Garden' (Figure 3.3) and 'The Cultural Park for Children' (Figure 3.4), were also developed in 1987 in the historic districts of Al-Fustat and Al-Sayyida Zeinab respectively. Simultaneously, the Aga Khan Trust for Culture donated to the Cairenes the largest project in its Historic Cities Programme, and the largest park in Cairo, 'Al-Azhar Park' on the Darrassa Hill in 1984, completed in May 2005 (Figure 3.5). These projects also involved the upgrade of a number of existing public spaces considered to be of significant value, e.g. historic, botanic, and so on, as well as the cleaning and

▶ **FIGURE 3.2**
The International
Garden

◀ **FIGURE 3.3**
Al-Fustat Garden (E.
Hassan 2007)

◀ **FIGURE 3.4**
The Cultural Park
for Children, 2007

◀ **FIGURE 3.5**
Al-Azhar Park

beautification of other public spaces present in the public's everyday life, such as squares and pavements (El-Messiri, 2004).

Formal space

This virtual walk thus continues with an exploration of the influence of the institutional working patterns and systems on the formal development of Cairo urban space. To adequately comprehend the institutional perspective within Cairo towards urban space, the main actors need to be identified together with their roles and strategies, as well as the hierarchy and relationships between them in the decision-making process. Accordingly, the following actors are identified: the Cairo government, the Administration of Cairo Transportation, Cairo Cleaning and Beautification Agency (CCBA) founded in 1983, and the Specialised Gardens Administration Project (SGAP) founded in 1987 (Cairo Government). Although we recognise the role of the Ministry of Agriculture within the local institutions in Cairo, it is not included in this presentation since its role is oriented more towards ongoing cultivation and maintenance of public space rather than the creation of public space.

First, the top-down institutional hierarchy within the Cairo government is recognised. The governor represents the highest rank and plays three main roles. The first emphasises the top-down approach within the government decision-making process and involves a direct impact on the development of public spaces in the city. These decisions vary from the establishment of new specialised departments to the initiation of area-based projects. El-Messiri (2004) illustrates how the 'appointment of a Governor who is an agriculturist' in this context of growing interest and attention towards the scarcity of green areas in the city helped to establish the Specialised Gardens Administration Project. The governor is thus responsible for the allocation of new gardens/projects to this institution. Simultaneously, the governor plays a secondary role in the decision-making process, through the processing and evaluation of suggestions and projects presented by other actors. Finally, the government office assesses and appraises the other involved actors; they design the guidelines for the roles and strategies of these actors using assessment criteria. For example, a competition was held in 2006 between the districts to choose the best district in each of the four regions.

In this context, the Specialised Gardens Administration Project and Cairo Cleaning and Beautification Agency represent the two main actors directly involved in the development of Cairo public and urban space. SGAP comprises 27 special gardens, i.e. gardens with historic, botanic, and/or artistic value and significance. Today, SGAP 'ranks at the top of the administrative hierarchy' (El-Messiri, 2004: 228). It helps to service the middle-class sector, who despite being unable to afford membership in private clubs and communities, refuse to use popular public spaces. Financially, it is a self-sufficient project receiving no government funds and its income is based on entrance fees, renting spaces to cafés, theatres, and other entertainment activities within the gardens (El-Messiri, 2004). The other, non-specialised gardens and public spaces, such as squares and sidewalks, are the responsibility of CCBA. CCBA is a government-funded agency whose responsibilities involve cleaning and waste disposable, planting, basic landscape and management of green areas, lighting, and maintenance. However, Administration of Cairo Transportation is often involved in the CCBA decision-making process. Urban projects are allocated to Administration of Cairo Transportation to plan and design, and hence an urban space, a certain square or a side walk for example, is ascribed to the authority of CCBA to clean, plant, and add lights. This decision-making hierarchy results in the preference for and dominance of vehicle design requirements over the needs of pedestrians and public life.

Accordingly, SGAP has an independent role towards special gardens, while the CCBA role is circumscribed by its relationship to Administration of Cairo Transportation. The division of responsibilities between SGAP and CCBA creates two (social) classes of public space: a higher class requiring entry fees, and a popular class providing free entry respectively. This separation is obviously reflected in the urban space in Cairo; Al-Fustat Park, which is built in an old and relatively poor district lies within the responsibilities of CCBA; whereas parks like the International Garden in Madinet Nasr, a relatively modern district, come under the auspices of SGAP. Although the decision-making process follows a top-down approach, it does not reflect a rigid hierarchal process, but the coordination between the actors involved needs to be revised. It must also be noted that the people involved in this decision-making process come from different backgrounds – agriculture, business, military, and police – but there is limited involvement of architects and urban designers. Finally, it is worth noting that these actors are more concerned with the development and increase in numbers of green areas rather than the design of public spaces.

A deconstruction reading of Cairo urban space

> On the one hand, there is almost uniformly a deeply felt social need to continually re-affirm traditional values, cultural, and even national identities. On the other hand, there a wholesale commitment, even infatuation, with modern western technology associated with participating in the geo-political economic order and in reckoning with the very real problems of rapid growth in urban population, largely occasioned by this participation.
>
> (Hakim and Rowe, 1983: 22)

The reading of Cairo space in both this chapter and 'A Threshold' associated her identity with many synonymous binaries in between the local and the western. The dominance of these binaries has hence created a chain of inner conflicts within her city space (Elsheshtawy, 2004). These binaries were approached through themes like orientalism and post-colonialism (Said, 2003). An orientalist image represents the oriental city as it developed in the past, which is enchanting but handicapped and frozen in time, unable to move on. And a post-colonial reading helps to relate the city to a 'narration of loss' of the glory of the tradition, the local, and so on through colonisation.

The question raised by or continuing through this reading is: 'could "deconstruction" help the perception of the city to go beyond the inherited binaries, and the "monolithic" representation?' Or more appropriately 'could "deconstruction" help to reveal the hidden and inherited cornerstone to deconstruct these binary readings and representations through her space and routes?' Deconstruction aims to show the limitations of a binary representation of reality rather than its elimination. Thus, deconstruction is not interested in the binaries which are reflected through a living reality but in the binary representation of unjustified play of power in space; see Figure 3.6.

▶ **FIGURE 3.6**
A reading instance in the binary representation of Cairo space

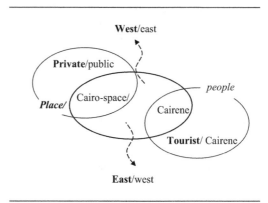

I thus attempt to identify the binaries inherited in the city and the display of power beyond these representations. The aim here is highlight the cornerstone that helped to associate the city to a monolithic 'oriental' image and identity (Abdelwahab, 2013). This helps to open the way to a reading deconstruction event that appreciates the dynamic realities and complexities in the city urban space. This deconstruction reading of Cairo space thus considers the setting of the reading space, meaning, and event as introduced in the previous chapter.

Beyond singular/in-context reading

The reading of Cairo and Egypt in general has always been associated with the context outside her space. This led to the dominance of the historic context, and particularly of specific chapters in her history. Consequently, a singular reading of Cairo acknowledges her relation to both eastern and western cultures beyond any binary opposition that implies the dominance of one term and subversion of another (Figure 3.7). The deconstruction of the supremacy of this binary in her context allows the development of a new reading that walks through her space to discover the multiplicities and dynamics within. These discussions highlight the blurriness of the boundaries between urban space, in the case of this study for example the space of Cairo or Egypt, and the outside context, whether global, western, regional, temporal, and so on. This context is simultaneously inscribed inside this urban space. Furthermore, the history of her space is read through that

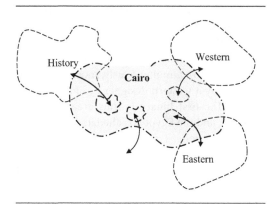

space rather than being dominating by it; i.e. it is possible to acknowledge and appreciate the different chapters of history within her space without approaching the reading of this space through a historic or traditional frame of reference. It should be noted that no 'single' reading is anticipated but multiple, changing interpretations on the way to uncovering the meanings, identities of her space.

A city of aliens/foreigners

[A] history [that] [began] when a particular 'us', who are not 'them' . . . is the ambiguous struggling through and with colonial pasts in mak-ing different futures.

(Verran, 2001: 38)

The social space in Cairo is primarily concerned with the binary us/them: public/private, citizen/foreigner, local/eastern, eastern/western, etc. (see Figure 3.8). These binaries are mostly developed and represented through a contextual reading of Cairo, which I attempt now to re-approach through a deconstruction singular reading to identify the display of power beyond these binaries. At the centre of this reading lies the fact that Cairo was

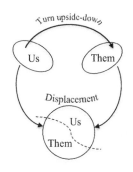

Searching for conflict

built by foreigners for foreigners excluding the Egyptians outside her space. 'Fatimid Cairo was a "forbidden city" . . . enclosed within heavy and gated walls . . . reserved for the Caliph. . . . [Whereas] the Egyptians . . . worked in Cairo . . . but when they finished their duties, they were expected to leave' (Thompson, 2008: 174–175). Simultaneously, modern Cairo is 'in short . . . a western quarter . . . by a western designer . . . for a western public' (El-Messiri, 2004: 224). The first binary is thus created between these two spheres: the foreigners and the Egyptians alienated from their home.

On the margin, the foreigner's cities showed tolerance and acceptance to other cultures (Raymond, 2007). The Islamic city of Cairo became the destinations of scholars from different countries including Egypt; the Egyptians were given a chance to reclaim their way back to the city, being scholars themselves. Simultaneously, the modern city was a residence to many western cultures including French and English, among others, and 'also shared by the elite Egyptian public who aspired to a western way of life' (El-Messiri, 2004: 224). Accordingly, another sphere was created in between the other two – foreigners/aliens, them/us – that involved scholars of religion in the Islamic city displaced by westernised 'elites' in the modern city. The latter are educated in secular Egyptian universities like Cairo University rather than Al-Azhar religious education, and some followed a western education.

Accordingly, we have two states founded by foreigners (the Arab later displaced by Turks and westerners), for foreigners that alienated the Egyptians, particularly the religious sector, both physically and socially outside the gates of the elite space. The Egyptians thus created a third sphere that claimed intellectual identity (religious later displaced by secular) to deconstruct the boundaries between the foreigners inside the city space and the aliens outside. It should be noted here that these displacements did not replace each other but continued to exist displaced in the city creating a conflict over the ownership of the city space; see Figure 3.9. This struggle was manifest through a history of revolutions and resistance, e.g. the 1919 revolution. Furthermore, despite the similarities between the two cities, the modern city created yet another layer of alienation of the Egyptian identity that has grown over almost 900 years in attachment to the Arab identity, the other foreigner.

> [I]n Ismaíl's urban projects of creating two cities side by side intensified. Before 1882, the dividing line separated a traditional sector from a modern one, but [in the British city] . . . the line marked a boundary between different nationalities, a harsher and more intolerable division. One could now speak of a native city and a European one.
>
> (Raymond, 2007: 333)

An orientalist image, as defined by Edward Said (2003), represents the oriental city as it developed in the past, which is enchanting but handicapped and frozen in time, unable to move on; and brings about a 'narration of loss' of the glory of the tradition, the local, etc. through colonisation. 'Colonialism . . . is oppression and destruction of the indigenous forms of knowledge by powerful Western forms of knowledge' (Verran, 2001: 26). Western representations are hence considered as parasitic additions to the city space; the

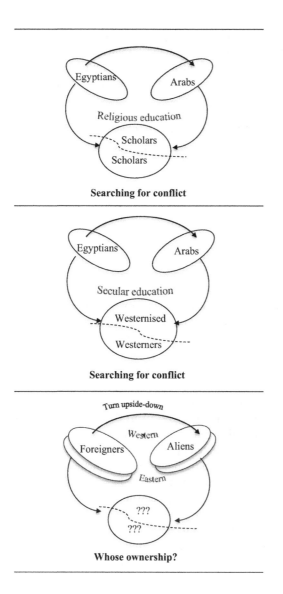

◀ **FIGURE 3.9**
Cairo-
deconstruction
space (2)

local requires 'the expulsion of invading western forms by a renaissance and resurgence of indigenous forms of knowledge' (Verran, 2001: 27).

These two readings, orientalism and post-colonialism, are in reality, attempting to understand and go beyond the monolithic representation of the city through the image of the historic city. They challenge this monolithic image through a demonstration of its dependency on the western binary west/east. The western binary privileges the west over the east, and hence values the oriental and pre-colonial image that contrasts the western. However, we need to highlight that these readings have to avoid only reversing the western binaries in the context of Egypt; the eastern is privileged over the western, which is controversially attached to the privileged western binary. The reversed binaries east/west, oriental/international are concurrently displaced, to be deconstructed. Simultaneously, the inner polarisation of the city between the modern and the traditional is another contextual binary, which allocates her identity in relation to either the western world or the regional area.

Nostalgia/aspiration for a conservative novelty

> What could be the origin of the feeling of being alienated without leaving home?
>
> (Amin, 2011: 133)

'On the 26 July [1952], Farouk [and the Turkish royal family] sailed from Alexandria' and he left the country (Thompson, 2008: 292). This was a first step in the radical change in Egypt towards foreigners and foreign policy, giving way to the 'national project'. The national project mainly sought to liberate Egypt and her economy from foreign dominance and interference (Amin, 2011). And 'On 26 July 1956', the nationalisation of the Suez Canal started a series of project nationalisations, which, together with Egypt's influence on the Arab-western relations, attributed towards a conflict between Egypt and the west (Amin, 2011; Thompson, 2008). This conflict was manifested in 1956 through the 'war of the tripartite': British, French, and Israeli, on Egypt. Following this 'citizens of those countries . . . were ordered to . . . depart immediately, [and] . . . were not allowed to return' (Thompson, 2008: 299). The conflict continued through the 'western economic boycott of Egypt' which 'prolonged her economic crises' through 1965–1966 (Thompson, 2008: 309). At the same, the economic crisis drove many Egyptians, particularly workers, to immigrate abroad which was facilitated by an 'ease [of] migration controls . . . that continues to the present' (Thompson, 2008: 309). Furthermore, Amin (2011: 140) highlights another phenomenon where Egyptians favoured self-alienation outside Egypt rather than 'living [alienated] in their own country' that grew with the development of the presidential city.

Amidst this turmoil, Amin (2011: 33) reflects on the Egyptians' 'powerful sense of belonging . . . up until the military defeat of 1967'. He also argues that a 'sense of alienation' of the Egyptians towards Egypt was created by 'the beginning of the 1952 revolution' and the formation of the 'military' city.

However, we have reflected on the alienation of Egyptians both physically and socially manifested in the capital city Cairo, precedent to this revolution. The argument demonstrated here is that the attempt to displace the modern 'western' city has alienated the 'Egyptian' identity associated with this time and space. This is also similar to the modern city's alienation of the preceding 'Egyptian' identity associated to the Islamic city. At the same time, the struggle of the developed new space after the revolution to reclaim the ownership of Egypt through rapid changes and replacements displaced the 'past' identities outside this new space, and then aspired to reclaim a new identity through the 'past' history of Egypt. Furthermore, the changing economic and social status displaced more Egyptians outside the country space. On the margin, the culture of tolerance towards the others, the foreigners, that has previously extended to allow the Egyptians back inside Cairo, was displaced by an intolerance that extended to push Egyptians outside, and further alienated them inside, both physically and socially (Figure 3.10).

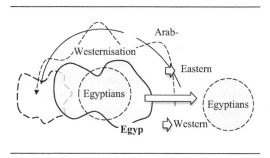

◀ **FIGURE 3.10**
The oriental representation: eastern/western binary displaced to the margins of Egypt space to highlight the Egyptian (people and place) ownership and sense of belonging conflict at the centre

The displacement in the discourse on Egyptian identity beyond the eastern/western representations towards a conflict of ownership and belonging is thus evident. However, the failure to recognise this displacement is also evident, through continuous misrepresentation. Singerman (2009: 4) argues that the 'national project' was the by-product of 'the state socialism and postcolonial nationalism of [Abdel] Nasser'. It intrinsically extended to Arab-regionalism; however it was 'also closely linked to domestic concerns' (Thompson, 2008: 295). The nationalist project helped to open the closed 'elite' public spaces: squares and parks, to the Egyptian public. However, it also marginalised the Egyptian identity in favour of Arab-nationalism in recognition of the past association of the two identities for a decade Amin (2011: 137). A second 'wave of westernisation' was then introduced by Sadat, which re-opened Egypt to external influences, 'often with crudity not at all agreeable to the tastes of a wide variety of Egyptians' Amin (2011: 140). This wave also brought a second wave of alienation of Egyptians from public life through the privatisation of public spaces: parks, palatial gardens, and Nile promenades. However, it should be noted that the 'past' British westernisation has been displaced by an American one (Amin, 2011; Thompson, 2008). Thus, the re-emergence of the Egyptian neo-identities associated with the Arab – eastern – and western presence in Egypt's history is not a continuation of the binary manifestation in her urban space.

The neo-identities are rather nostalgic to past identities and resistant to the continuous alienation of the 'other' identities. Accordingly, both representations continued in her space but also continued 'to decline and erode' (Amin, 2011), giving way to a westernised traditional image, in modernised Islamic architecture style.

WHO IS KHŌRA?

What is place? To what and to whom does it [Khōra] give place? What takes place under these names? Who are you, Khōra?

(Derrida, 1995: 111)

The word Khōra itself refers to 'place, location, region, county', as well as reflects metaphors proposed by Plato in *Timaeus*, 'mother, nurse, receptacle, and imprint-bearer' (Derrida, 1995: 93). To investigate Khōra and these complementary synonymies, two different readings related to architecture are presented in this section. The first explores Jacques Derrida's deconstructionist reading of Plato's 'chóra', presented through his unfinished essay 'Chóra' in 'Choral works', and later 'Khōra' in 'On the Name' (Derrida, 1997a; Derrida, 1995), and 'Plato's Pharmacy' in *Dissemination* where Derrida explored 'chóra' for the first time (Bennington and Derrida, 1993; Derrida, 1981). Another reading is represented through 'Chóra: The Space of Architectural Representation' in (Pérez-Gómez and Parcell, 1994). This reading was developed through an architectural interpretative background, 'We are now in a better position to understand the nature of chóra as paradigmatic architectural work' (Pérez-Gómez, 1994: 15). Derrida's reading on the other hand, 'is neither architectural in itself nor is it devoid of architectural relevance' (Grosz, 1995: 117); 'yet I have always had the feeling of being an architect, in a way, when I am writing' (Derrida, 1997a: 8).

One main difference between the two readings is that Derrida's reading of Plato, Khōra, remains distinct from Aristotle's place, topos (Rickert, 2007), whereas Pérez-Gómez's reading of chóra reflects Aristotle's reading of Plato. Pérez-Gómez considers that 'the work of architecture as chóra is indeed' related to/reflected in Aristotle's place, topos, space-matter, contained space, and material container (Pérez-Gómez, 1994: 30). As discussed in Chapter 1, Aristotle's concept of place 'subsumed chóra under "topos" and theorized it as material space' (Rickert, 2007: 253). Another difference lies in the names Khōra/chóra used by Derrida and Pérez-Gómez respectively; while both protect the Greek word from translation – Khōra 'remains caught in networks of interpretation' (Derrida, 1995: 93). Khōra represents a feminine given name/noun (Dutoit, 1995). Furthermore, Derrida's reading of Plato's *Timaeus* also reflects the writings of Socrates, Hegel, and Heidegger.

Derrida thus emphasised that there is no single definition of Khōra, who should not be considered as homogenous unit, although she is in fact a unit; 'there is only one Khōra, and that is indeed how we understand it; there is only one, however, divisible it be' (Derrida, 1995: 97). His sincere exploration of Plato's text 'offers a countersignature to that text' (Wolfreys, 2007: 71). However, post-readings of Derrida's *Khōra* have tended to emphasise a partial translation of the word; for example, the feminine and the place in between is referred to in Grosz (1995, 2001) and Bennington and Derrida (1993). Pérez-Gómez's reading, on the other hand, considered the Greek myth, both in Aristotle's contribution as well as those of Descartes and Galileo, to be a reflection of the development of architectural 'chóra' – in the west – since the time of the renaissance and the baroque.

Hence, I share Derrida's quest to understand 'Khōra', who 'seems never to let itself be reached or touched' (Derrida, 1995: 95). Khōra's names and metaphors are identified and studied as introduced through Derrida's reading and complemented by that of Pérez-Gómez as an oscillation, a feminine figure, a space, an impossible place, and a new paradigm. However, the interrelations and dependence of these names and metaphors are also taken into consideration, where each folds on to and helps to explain the others. However, each part would be a reflection on a partition or a characteristic of 'Khōra'. The word has two spellings: 'chóra' is the general use and employed by Pérez-Gómez; 'Khōra' on the other hand, is used by Derrida to express her feminine identity, as previously explained, and will therefore be used to refer to this entity.

An oscillation

> [Khōra] oscillates between two types of oscillation: the double exclusion (neither/nor) and the participation (both this & that).
>
> (Derrida, 1997a: 15)

This oscillation exists between two oscillating types rather than between oscillating figures. These two types are double exclusion (neither/nor) and participation (both/and). This oscillation denies polarity and binary opposition, which is arguably the basis of western thought and philosophy (Derrida, 1995; Grosz, 2001; Wolfreys, 2007). These binaries were set by Plato in *Timaeus*, and represented ordered hierarchal opposition between being/becoming, intelligible/sensible, ideal/material, divine/mortal, perfect/imperfect, homogenous/heterogeneous, and so on (Derrida, 1995; Grosz, 1995). These binaries are problematised through 'chóra', 'the passage [involves space and movement, an oscillation] from the perfect to the imperfect' (Grosz, 1995: 114), from intelligible to sensible. Accordingly, we shall explore the nature as well as the figures of this oscillation.

Nature of oscillation

The either/or relation considers a predetermined inclusion and choice between the two binaries; it has to be either this or that (Casey, 1997). The oscillation between the two oscillating types refuses determination, the well-defined black and white. While the double exclusion alienates Khōra from the binary couples, she does not belong to either; the participation problematically relates her back to each of the oppositions (Derrida, 1995). Khōra is therefore neither intelligible, an object of thought, nor visible to the sensible world (Derrida, 1995: 90). These definitions are also temporal, 'at times the Khōra appears to be neither this nor that, at times both this and that' (Derrida, 1995: 89). Accordingly the passage is 'at once of place and from place' (Casey, 1997: 45), that is Khōra 'the space in which place is made possible' (Grosz, 1995: 116) 'but it takes place only in place' (Casey, 1997: 45).

Figures of oscillation

As discussed, Khōra oscillates in *Timaeus* between the intelligible and the sensible, mind and body (Grosz, 1995). The intelligible, the object of thought, which is understood by the intellect (mind), is an eternal model (being), ideal, divine, and not apprehended by sight or senses. The sensible, on the other hand, is apprehended by the senses (body), is in the process of (becoming), constantly in motion, material, mortal, and imperfect, the being/thought that 'has come into existence' (Derrida, 1995; Broadbent, 1991b; Grosz, 1995; Pérez-Gómez, 1994: 8; Casey, 1997: 35). And, the 'everlasting' Khōra questions the distinction between the sensible and the intelligible (Derrida, 1995; Rickert, 2007). Plato also perceives Khōra as 'space and the condition for existence of material objects', and yet 'is apprehended without senses' (Grosz, 2001: 91; Pérez-Gómez, 1994: 8). Simultaneously, Khōra participates in 'a very troublesome' way in the intelligible world (Derrida, 1995) (Figure 3.11).

▶ **FIGURE 3.11**
Khōra, figures of
oscillation

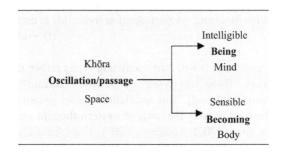

A genre: is she feminine?
=======================

> [Khōra], which is neither 'sensible' nor 'intelligible', belongs to a 'third genus'.
>
> (Derrida, 1997a: 15)

'The discourse on Khōra is also a discourse on genre/type' (Derrida, 1995: 91). Simultaneously, Derrida problematises the definition of 'the passage', explored in the previous section, between all these binaries (both/and) and a third genre 'another which would not even be their other' (neither/nor) (Derrida, 1995: 104). This double oscillation and this third genre deny any distinction; 'Khōra . . . names the deconstruction of, and taking place between genders . . . genres and the idea of genus also' (Wolfreys, 2007: 73). This third genre is associated with the mother, nurse, receptacle, imprint, etc. (Bennington and Derrida, 1993; Derrida, 1995). It also relates problematically to the feminine gender. Khōra is a 'feminine noun', which Derrida replaces with a feminine pronoun 'elle', or 'she' (Dutoit, 1995: xii).

> Khōra is not more of a mother than a nurse, is no more than a woman. . . . She does not belong to the race of women.
>
> (Derrida, 1995: 124)

Derrida hence problematises Khōra's association with a mother, a nurse, a feminine character. 'Taking the risk of saying that it's called "mother" is also to recognize that one no longer has a very clear idea of what a mother is' (Bennington and Derrida, 1993: 205). A mother is not the binary opposite of the father but a third genre in between (Bennington and Derrida, 1993; Derrida, 1995). In similar projection, Khōra is not a feminine, as opposed a masculine figure, but a third genre. 'She engenders nothing and besides, possesses no property at all . . . nor ownership of children' (Derrida, 1995: 105).

> Khōra is not . . . anything but a support or a subject which would give place by receiving or by conceiving, or indeed by letting itself be conceived.
>
> (Derrida, 1995: 95)

Simultaneously, Khōra is the reciprocal, which designates her function towards the intelligible, the sensible as well as towards all the binaries, to receive, conceive, and give to the world (Derrida, 1995; Grosz, 1995) (Figure 3.12). However, Khōra inherits no qualities for herself; 'she possesses them, she has them, since she receives them, but she does not possess them as properties, she does not possess anything as her own . . . nevertheless, she is not reducible to them' (Derrida, 1995: 99). Accordingly, to assign any 'particular property' to Khōra would lose her status as a reciprocal (Grosz, 1995: 114).

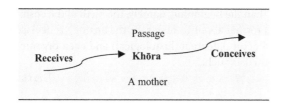

◀ **FIGURE 3.12**
Khōra the reciprocal

Derrida reads Khōra as a feminist figure but she does not possess the properties of the feminine gender. A feminist approach could be pursued therefore through Derrida's writings; Julia Kristeva (1984 [1974]) and Luce Irigaray (1985) are key figures in the feminist reading of Khōra (Casey, 1997; Grosz, 1995; Rickert, 2007). Simultaneously, Grosz (1995: 113) 'reading Plato and Derrida on chóra', argues that 'the notion of chóra serves to produce a founding concept of femininity'.

An architecture space – an impossible space

> [Platonic] chóra is both cosmic place, and abstract space, and it is also the substance of the human crafts.
>
> (Pérez-Gómez, 1994: 9)

This quotation helps a reflection on three readings of architecture space through chóra. Derrida (1995) approached architecture space as a realisation, the practice of Khōra, as in Plato's chóra. Accordingly, he asked the architect to build Khōra, the impossible place. For Pérez-Gómez (1994: 9), on the other hand, architecture space is a reading of Aristotle's topos which subverted chóra to matter, 'the substance of the human crafts'. Interestingly, Derrida rejects the association of Khōra with matter, 'a word Plato never used to qualify Khōra' (Derrida, 1995: 127). Berque (2005) rejects both readings; he combines both chóra and topos in a concept of architecture space, which he approaches as a relationship between people and place. Thus, he takes the idea of chóra even further, into the 'milieu', the landscape, the urban context (Berque, 2000).

Simultaneously, according to Plato, Khōra, which is almost always associated with a trilogy, is combined with the intelligible and sensible to represent reality (Pérez-Gómez, 1994; Casey, 1997). Khōra 'is the place always on the move' in between the sensible and the intelligible (Lucy, 2004: 68); in between space and place, in between architecture space and landscape (urban context). Accordingly, 'chóra is granted a strangely displaced place', which is hence associated with the beginning, the cosmos (Derrida, 1995; Lucy, 2004; Rickert, 2007: 256), and whose association with architecture is simultaneously complicated (Grosz, 2001; Rickert, 2007).

Accordingly, I will hence consider the formal analogy of Khōra as a spatial figure, her extension to the cosmos and the context, and finally, her displacement as an intermediate space, an impossible architectural space.

An analogy: a spatial figure – thinking relations

> In order to think Khōra, it is necessary to go back to a beginning that is older than the beginning, namely, the birth of the cosmos.... In that which is formal about it, precisely, the analogy is declared: a concern for architectural, textual (histological) and even organic composition is presented as such.
>
> (Derrida, 1995: 126)

Khōra 'seems never to let itself be reached or touched' (Derrida, 1995: 95). In between the etymology of the word as space/place, location, region, and so on, and such figures of speech as nurse, receptacle, architect, and so on, Khōra is 'caught in networks of interpretation' (Bennington and Derrida, 1993: 93), networks of complex structures, 'radically different types of structures', inscribed within 'a chain of relations where there is no relation', but non-similar analogies that go beyond formal identities (Wolfreys, 2007: 72–73). Thus, the formal analogy of Khōra as architecture is complemented in Derrida's former quotation by histology – 'A branch of anatomy concerned with the study of the microscopic structures of animal and plant tissue' (Encarta-Online-Dictionary) – and organic composition. As previously discussed, Khōra is on the move; as soon as her definition is read, her definition shifts.

The cosmic: the milieu – urban context

> [Khōra] . . . includes the sense of political place, or more generally, of invested place, by opposition to abstract space. Khōra 'means': place occupied by someone, country, inhabited place, marked place, rank, post, assigned position, territory, or region. And in fact, Khōra will always already be occupied, invested, even as a general place, and even when it is distinguished from everything that takes place in it.
>
> (Derrida, 1995: 109)

The motion of Khōra leads to the arché, 'as generally in architecture, architectonics, arche-écriture or arche-writing' (Wolfreys, 2007: 72); the arché, the beginning, the cosmos. Pérez-Gomez (1994: 16) takes the analogy further, to find 'the ever-present origin of . . . architecture' in space, in place, in chóra, and in Khōra, a spatial figure, 'in which becoming happens . . . is indeed milieu', a spatial milieu (Berque, 2000: 7). This analogy leads to the urban context 'with all its geographical and social particularity' (Miller, 2001: 165), inhabited, political . . . landscape. But again Khōra is 'the secret place-without-place hidden in every topography' (Miller, 2001: 93–94).

In between: an intermediate space

> The in-between is . . . inimical [hostile] to the project of architecture as a whole.
>
> (Grosz, 2001: 94)

Grosz (2001: 91) argues that Khōra's lack of 'a fundamental identity . . . a form', through her 'position of the in between', is hostile to architecture and urban space, which is intrinsically concerned with form and identity. Khōra, the reciprocal, the passage, and the conceiver, inhabits an in-between space, through a double oscillation in between two genres. Thus, Khōra is a third genre that goes beyond other genres and types. She loses her distinction, determination, and definition, as she deconstructs the 'polarity' of western metaphysics (Derrida, 1995: 92). However, the space in-between does not actually exist, as it continually oscillates in between other spaces. 'There is Khōra but Khōra does not exist' (Derrida, 1995: 97). Khōra is beyond category, and categorisation. She receives her identity from other spaces, to pass on and conceive other spaces but not to keep them.

> The space of the in-between is that which is not a space, a space without boundaries of its own, which takes on and receives itself, its form, from the outside, which is not its outside . . . but whose form is the outside of the identity, not just of an other . . . but of others.
>
> (Grosz, 2001: 91)

A paradigm shift

Finally, we shall review how the recent reversion to Plato's chóra has produced a paradigm shift in the reading of space/place. The development of 'space/place' in the history of philosophy since Plato's chóra was reviewed in Chapter 1, together with its resurfacing in contemporary philosophical considerations of place through the works of Julia Kristeva (1984 [1974]), Jacques Derrida, and Gregory Ulmer (1994) (Rickert, 2007: 252). This emerging interest in space – chóra – does not entitle it to a fixed definition but is rather an attempt to find the dynamics of space/place 'at work' (Rickert, 2007). Again it was Derrida, Eisenman, and Tschumi who addressed these concepts in architecture (Casey, 1997: 286). As previously discussed, this book draws on a special interest in Derrida's architectural experience 'Khōra'. Thus, interest in neo-Platonic chóra is augmented through this collaboration with architecture.

Rickert (2007) goes on to discuss how this interest in 'Platonic chóra' produced a new paradigm shift towards the concept of space/place, changing the previous understanding about the relations and interactions of the body with the spatial environment. Space/place constitutes mind, body, and space-matter. The traditional 'separatist' paradigm held these as autonomous independent spheres. The mind is rational, an intelligible logic. The interrelations between these spheres follow a direct and linear method, an organised internal structure. There is a plan, a method to achieve this plan, a spatial arrangement. Accordingly, the traditional paradigm is represented in Figure 3.13, through three independent spheres; the mind interrelates with body and space through a linear and direct path.

▶ **FIGURE 3.13**
Traditional paradigm of space/place

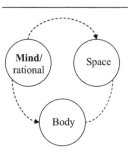

Rickert (2007) traces a new paradigm of place which challenges the traditional understanding of body and space relation and interaction. Mind, body, and space are no longer autonomous. Also, the modern separatist notion of space/place has initiated the desire to revert back to the ancient holistic approach through the emphasis on mind and body. The mind is both 'emotional and rational' questioning the relations between the self and the world. It is 'embodied and dispersed in' body and space (Rickert, 2007: 251). At the same time, it is also immersed in its social and technological contexts, as demonstrated through the review of Tschumi's reading in the previous chapter. The interrelation between these spheres no longer follows a linear method but consists of multiple and complex negotiating systems (Rickert, 2007). It was also shown that the new paradigm shows a particular affinity towards architecture. Place/space is an architectural component, a medium for thoughts and actions, activities, the interaction between the self and the other, as well as between bodies/people and the spatial environment. The new paradigm is outlined in Figure 3.14. Finally, Rickert also reflects on

▶ **FIGURE 3.14**
New paradigm of space/place

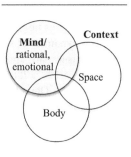

the diminishing boundaries between inside/outside through this complex, dynamic notion: Contemporary work on the chóra suggests that there is no clear demarcation of 'in here' and 'out there', and that the notion of system is not one of directly following a method, in some linear fashion, but being immersed in, negotiating, and harnessing complex ecologies of systems and information (Rickert, 2007: 253).

A DECONSTRUCTION READING OF URBAN SPACE

> Derrida [is] always [concerned], with the place of Egypt (or more accurately the figure of Egypt) as an origin of western philosophy.
>
> (McQuillan, 2010: 255)

Cairo urban space, as elaborated in this chapter, is primarily concerned with the binary 'us who are not them'. An orientalist (exotic) reading imposed on the city scape that in its continuous rejection of 'them' is whole heartedly accepting the binary of us/them, west/east. And though this binary of the modern double city was displaced in the contemporary complex city representation, this displacement remained unrecognised and accordingly remained at the centre of the city reading space. Significantly, city reading space is thus a representation space of the deconstruction project as introduced by Derrida in search of the origin of western metaphysics: west/east. Accordingly, the question raised by this chapter is 'could deconstruction help in the perception of the city to go beyond the inherited binaries and monolithic representations?' And the reply would be that the old patterns and static relations are becoming exposed, and recognition of the false 'monolithic representation' of the city space is highlighted. However, I would agree with McQuillan (2010: 276) that 'even now posing the question of Egypt is only in its infancy'.

This reading also highlights the need to involve a vignette from Cairo in the context of this book. An essential criterion is the involvement of an architect or urban designer in the development of the project. At the same time, the size and location of the project is important. These criteria suggest the inclusion of 'The Cultural Park for Children' built in the Al-Sayyida Zeinab historic district. This park designed by the Egyptian architect Abdelhalim Ibrahim is rather a small project – in size – but, it won both a national and international architectural award. Accordingly, this chapter proposes to re-approach this vignette in Part II.

> ~~Place~~ [Urban space] would indeed be a less misleading translation of ~~chóra~~ [Khōra] than space.
>
> (Casey, 1997: 353)

Simultaneously, Derrida protected Khōra from translation and interpretation through language, keeping her proper name and feminine figure, without pursuing a feminist position. However, in his quest to build Khōra – refer to Chapter 2 – he subjected her to a different kind of interpretation, an architectural one. Khōra, which is as ambiguous and abstract as an architecture space of knowledge, is forced into the material/physical space and form.

This approach to reading Khōra is an attempt to displace her again in between idea and reality, intelligible and sensible. As demonstrated in Chapter 1, architecture space is a space of knowledge (intelligible concept) of the physical space (sensible form), immersed in context (Figure 3.15). Accordingly, we reflected on traditional thinking about the transcendence of the outside space of knowledge to the immanent architecture practice and hence, physical space. This discussion brings about three simultaneous in-between spaces. Khōra is the space in between the intelligible and the sensible. She oscillates between these spaces, as she receives from a transcendent intelligible to conceive the sensible. Tschumi, on the other hand, displaced the modern dichotomy, 'form follows function' through the introduction of an intermediate space, 'concept-form'. This 'concept-form' space thus reflects the neutrality of the architecture space container that does not follow a function but accommodates any function. However, Khōra is also the deconstruction of the container space, as it blurs the boundaries between the binary 'contained space and material container' (Pérez-Gómez, 1994: 9).

▶ **FIGURE 3.15**
Khōra, a space
in between
architecture space

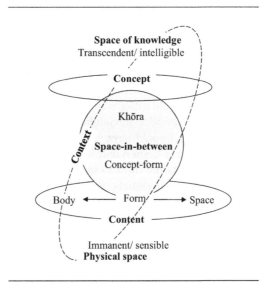

This reading – the oscillation of Khōra – is, accordingly, turned over through architecture space. In architecture, 'content and context are given elsewhere outside architecture', while the concept is immanent in the practice, 'it is what I, as an architect, have to generate' (Tschumi in Walker, 2006: 166). 'Nevertheless it is somehow contaminated . . . by context and content'. This contamination brings about the third space in between, which was also discussed in Chapter 1, in between transcendence and immanence. The transcendent intelligible outside knowledge which is only approachable through an immanence, appears to be inside the boundaries of the container space, transcending from inside the practice i.e. a space in between which contaminates architecture space, concept, content, and context.

At the same time, Chapter 1 presented a view in between the intelligible and the sensible, a world of subverted realities, namely pragmatics, inspiration, and creativity, and also reflected on the association of these realities with architecture space. Again, both architecture space that is not limited to the knowledge of form, and Khōra resonate as if they co-exist in parallel worlds. However, architecture is the art of place-making; and Khōra is the space without space that could only be approached 'as if in a dream' (Derrida, 1995: 99).

Consequently, my argument is that Khōra, the impossible place, is not hostile to architecture space as argued by Grosz (2001), nor can architecture build Khōra as requested by Derrida. Grosz's (1995: 116) argument is based on reading chóra as a 'the space in which place is made possible'. Khōra inhabits architecture and urban space, as she receives from the architectural space of knowledge embedded in the content and context of urban space, to conceive this urban space in content and context. A logic that could only be apprehended as Khōra oscillates between oscillations in the impossible place. Again, 'there is only one Khōra, and that is indeed how we stand it, there is only one, however divisible it be' (Derrida, 1995: 97). Khōra inhabits architecture/urban space as architecture/urban space inhabits Khōra; accordingly they are both the abstraction and realisation of each other through place, without building that place.

> Derrida's chóra inhabits an impossible place, one that governs, in a manner nearly meta-metaphysical . . . and the conflicts that emerge with Eisenman stem from Derrida's attempt to realize this impossibility leavened with an intuition that it cannot be realized – that it remains impossible.
>
> (Rickert, 2007: 266)

A TURNING POINT

This book aims to develop a ~~multidisciplinary~~ [integral/reflexive] ~~framework~~ [reading] of urban space, with special reference to Cairo-space, that operates ~~on~~ [between] ~~two~~ [multiple] ~~levels~~ [projections] of ~~urban space/place and context~~ concept: space/place, content: experiential place, and context: social, economic, political and so on.

(A Threshold)

This interlude represents a turning point to revisit the reading of urban space in response to the preliminary reading in Part I. It thus questions the development of the main aim: a framework to approach urban space. The preliminary reading intrinsically explored the multitransdisciplinary nature of the various approaches to urban space, both empirical and theoretical. A primary consideration of this multidisciplinary reading, as discussed in the threshold, is the need to develop it into an integrated report, rather than three or two parallel approaches; into an architectural theory of urban space in relation to post-structuralist philosophy and deconstruction on the one hand, and to social studies on the other, as well as including the interrelation between post-structuralist and social studies. The preliminary reading thus reflected on the similarities and differences between these disciplines, how they are integrated and how they interact so that it was particularly hard to define distinct paths in each approach. As a primary step towards dissolving the boundaries between these disciplines, I developed a primary framework of urban space, which is expected to work on two separate levels: space/place – in between idea and reality – and context – social, economic, political, and so on.

The use of a 'framework' implies a consistent analysis of a well-defined urban space; however I have also demonstrated that urban space is not 'well-defined', either in the complex literature and theories or in the everyday realities. Simultaneously, the preliminary reading elaborated the changing paradigm of space/place drawing on the deconstruction reading of 'Khōra'. The old paradigm considered the constituents of space/place as separate elements with linear and direct relationships, which could be approached through a structured framework. The new paradigm, on the other hand, describes these constituents and internal relationships as disperse, complex and negotiable, thereby emphasising the inappropriateness of the term 'framework'. At the same time, the preliminary reading emphasised the difficulty in approaching the dynamics and multiplicities of urban space through separated levels, which imply hierarchal framework of analytical steps. Therefore, the new paradigm is rather a projection of the dynamics and complexities developed within the urban space without hierarchy or guidelines, and accordingly it allows an approach to the not-well-defined constituents and relationships: concept, content, context. The next step thus requires a careful, continuously reflective, study to manage the rich complex data construction in such a way that allows a continuous interplay between the various sets of data.

Consequently, I explore the notion of reflexivity which involves the capacity to approach this continuous interplay and avoids the dominance of a specific theory or approach. I thus adopt Alvesson and Skoldberg's (2000) quadri-hermeneutics example of reflexive interpretation, four levels of interplaying interpretation: data construction, primary interpretation, critical interpretation, and self-reflection. The main objective is to develop the primary framework of space introduced in Chapter 1 through this notion of reflexivity.

REFLEXIVITY

> [Reflexivity] . . . in the sense of [a] & [b] being like two mirrors facing each other and constantly and endlessly reflecting their images back and forth between each other.
>
> (Gee, 2004: 97)

What is reflexivity? According to Alvesson and Skoldberg, reflexivity entails continuous reflection between different levels of interpretation, interaction with empirical data, interpretations of underlying meaning, critical interpretations of ideology, power and social discussion, and self-reflections on text and authority (Alvesson and Skoldberg, 2000: 250). Reflexivity works through the belief that every piece of empirical material mirrors a piece of reality. Accordingly, it considers two processes, interpretation and reflection. The act of interpretation involves deep considerations of all 'trivial and non-trivial' data, primary and secondary data as equally important. And reflection is an act of 'interpretation of interpretation' through the embedding of these interpretations in their context, in the socio-cultural background of the author as well as in the linguistic, ideological, and political considerations of the subject. Accordingly, reflexivity promotes a conscious process or mechanism which could help the author develop critical self-reflection and awareness, and possibly lead to change. 'An important function of reflexive analysis is to expose the underlying assumptions on which arguments and stances are built' (Holland, 1999: 467). Thus, the main aim of reflexivity is to 'generate knowledge that opens up rather than closes' i.e. open-ended research which does not establish the truth of reality (Alvesson and Skoldberg, 2000: 5). Accordingly, it aims to understand this truth through the continuous acts of exploration rather than establishing a particular truth. At the same time, reflexivity presents a specific type of reflection. While reflection gives a 'focus' on a specific level of interpretation, reflexivity involves a continuous interplay between the various levels without any of them becoming dominant. It also involves the capacity to change these interpretative levels as well as among each other. Accordingly, it entitles 'breadth and variety' rather than determinant interpretation (Alvesson and Skoldberg, 2000).

I shall hence adopt Alvesson and Skoldberg's (2000) quadri-hermeneutics example of reflexive interpretation, comprising four levels of interplaying interpretation: data construction, primary interpretation, critical interpretation, and self-reflection. Data construction considers the empirical material from both theory and fieldwork. Primary interpretation represents the basic level of interpretation that questions the underlying meaning of the empirical material, and strives for multiplicity and plurality. Critical interpretation principally questions and challenges the prime interpretations. And finally, self-reflection questions the interpretation process through reflection on interpretation authority and selectivity.

Simultaneously, the reading of urban space is reviewed through the vignette of Cairo, Khōra – the new paradigm of space/place in philosophy – the different theories of urban space introduced in Chapter 1, and the developed primary framework, as well as a particular emphasis on deconstruction. The main aim is thus to develop 'reflexive reading strategies' of urban space. In this sense, the empirical data constitutes both the vignette of Cairo and the theories of urban space. To process this data, two levels of interpretation are needed. The first is a primary structured approach to interpret the 'underlying meaning' of the data. The 'primary framework' of urban space is thus used to interpret the constructed data. Accordingly, 'strategies of deconstruction' represent the critical interpretative level of questioning and problematising the primary interpretations. A self-reflection on the process of interpretations as well as the developed reading strategies is conducted through the reading of Khōra, theoretical, and Cairo space, empirical.

In summary, the reflexive approach in this book comprises four parts: data construction setting the reading space – both the empirical vignette of Cairo and the introduced theories of urban space; primary interpretation – the primary framework; critical interpretation – deconstruction strategies; and self-reflection – Cairo-Khōra, as shall be further discussed.

SETTING THE READING SPACE

This is the first level to approach the construction of the urban space discourse in this reading. This level begins in the threshold by setting up the book thesis and the preliminary reading. This level thus continues two parallel paths: the empirical discourse through the vignette of Cairo space and the theoretical discourse elaborated in the preliminary reading. These will then be further analysed and interpreted to lead to the development of reading strategies. Accordingly, this section explores both the definition of discourse and the construction of urban space discourse that consists of two parts: the theoretical and the empirical.

What is discourse?

Discourse is part of everyday communication. It is mostly used to refer to comprehensible 'talk, writing, and signage' (Johnstone, 2008). Discourse is:

> [W]ritten or spoken communication or debate; a formal written or spoken discussion of a topic.
> (Pearsall, 2001 – Oxford English Dictionary)

> Any coherent succession of sentences, spoken or (in most usage) written.
> (Matthews, 1997 – Oxford Dictionary of Linguistics)

> A continuous stretch of language containing more than one sentence: conversations, narratives, arguments, speeches.
> (Blackburn, 2008 – Oxford Dictionary of Philosophy)

The last definition echoes the book's interest in philosophy. Discourse is addressed through different areas of philosophy, analytic, rhetoric, language, linguistic, semantic, among others (Routledge, 2000). The former definition was developed through 'discourse analysis', which considers discourse within its social context and is hence concerned with 'talk and texts as parts of social practices' (Potter, 1996: 105). Another definition related to philosophy is introduced by Foucault, although this definition considers discourse in a 'different sense than in discourse analysis' (Alvesson and Skoldberg, 2000: 224). 'Analyst' discourse considers 'text' rather than linguistic units. Analyst discourse both influences and is influenced by its socio-cultural context, knowledge (Alvesson and Skoldberg, 2000; Georgakopoulou and Goutsos, 2004). Foucault on the other hand, is more concerned with technical and theoretical definitions of discourses; discourses are systems of production of knowledge as well as operative processes of power within the socio-cultural context (Mills, 2003; Wooffitt, 2005).

Actually, discourse is 'one of the most contradictory terms' in Foucault's works (Mills, 2003: 53). Foucault uses a different definition for discourse as 'the general domain of all statements, sometimes as an individualisable group of statements, and sometimes as a regulated practice that accounts for a certain number of statements' (Foucault, 1974: 80). Foucault's interest lies in the third definition of discourse which considers the rules and structures behind discourse production, rather than the discourse produced and the details of language use (Mills, 2003: 53; Alvesson and Skoldberg, 2000: 224). Discourse is therefore seen as a system that produces the objects and subjects of the discourse within its social and language context (Alvesson and Skoldberg, 2000). Furthermore, Foucault's discourse is built in close relation to power relations.

> Discourses are tactical elements or blocks operating in the field of force relations. . . . Force relations are characterised by dynamic and contingent, complex, power-full, strategic situations of heterogeneity, instability and tension, where systems of relations (such as economic, gender, knowledge and so on) confer. What counts as the 'truth' is thus a product of practices of relational power.
>
> (Foucault, 1981: 102)

Following on from this short review of discourse(s) definitions, the definition and use of discourse 'within' the context of this book is now discussed. Primarily both Analyst Discourse and Foucault's discourse involve all kinds of texts (Wooffitt, 2005): ordinary conversations, interviews, official government reports, academic writing, newspapers, and so on. Further, the definition of discourse purposefully varies through the different book sections. Initially, discourse is used in this book to refer to a group of statements about urban space: discourse(s) of urban space. These discourses are then reviewed to provide a background to help develop the primary framework. Accordingly, the approach to these discourses attempts to avoid theoretical preferences, for instance discourse analysis. The main aim is to explore data in advance of interpretation. Discourse definition develops by considering the structure and systems that produce urban space discourse through the interpretation of theories and vignettes in the following chapters.

Construction of urban space discourse

According to Foucault (1974), discourse is constructed through a social and cultural context and associated with institutions of 'knowledge and authority'. It requires a functional framework in order to construct the relations between the different discourse statements (Horrocks and Jevtic, 1997; Mills, 2003). Simultaneously, Nussbaum (2001) identifies the actors involved in the discourse, referencing literary work and music, as authors, narrators, readers, plus various other characters. She also emphasises the closely related roles of author and reader over the narrator and other characters. These actors are projected through architecture space; they can be identified as architect/urban designer/planner as (text) authors/(urban space) designers; the people using the urban space are hence the reader(s); whereas the press and other professional reports play the role of the narrator in this discourse. Other characters are involved such as administrators and managers as well as those from other technical areas. This projection of discourse actors on urban identified by Nussbaum (2001) is represented through the illustration in Figure TP.1.

Simultaneously, Vining and Stevens (1986) developed a user-based model to read/evaluate the quality of urban space, which 'explicitly' involves the user in this evaluation. It considers the relations between the urban space, designer, and user. The user relation starts with actions which affect the

► **FIGURE TP.1**
Actors of urban
space, a projection
through Nussbaum's
reading (2001)

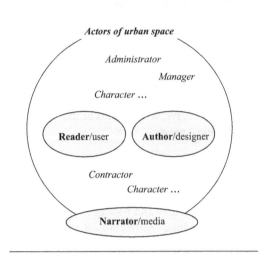

urban space and this changes the designer and manager perception of space capacities and constraints. Accordingly, the designer plays the role of communicator, informing the user about the consequences of their behaviour and thus affecting their future behaviour. The designer relation starts by altering the physical space. These changes are perceived and judged by the users, whose feedback is presented as demands and needs. In this way, they change the designers' perspectives of urban space.

The approaches of both Nussbaum (2001) and Vining and Stevens (1986) are thus adopted to identify the authors and actors involved in the urban space discourse; see Figure TP.2. This discourse is concerned with the relation between the idea and reality, the architectural concept and the product. It explores the relation between reading and interpretation of urban space on the one hand, and its construction on the other. However, it is not concerned with the production process. It is only concerned with the development of the idea and the end-product. In this context, it is appropriate to adopt Nussbaum's emphasis on the author/reader role in the construction of discourse, and the commentator on the periphery of the discourse, i.e. not directly involved in the production of urban space. The projection of these actors in Vining and Stevens's model helps to identify the author and reader/observer roles, the designer whose reading helped their intervention. These two roles are not contradictory, for the author is intrinsically the first reader of his/her writings. The narrator's indirect role is added and relations are built up in between the narrator and both the author and reader/observer. However, the user's direct role as reader/observer of urban space is excluded from the developed discourse. This exclusion of the user from the discourse setting helps to build up the roles of the designer as both author and reader of urban space.

Following this projection, the main authors are recognised in the vignette of Cairo space in Chapter 4, between the designer and the manager of The Cultural Park for Children. The secondary authors were identified through the press commentators and award committee. In Chapter 5, 'Social Space',

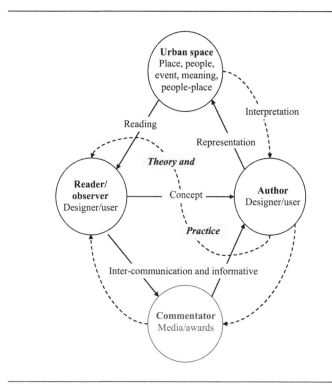

◀ **FIGURE TP.2**
Building up the institutional discourse based on a reading of Nussbaum, (2001) and Vining and Stevens (1986)

these authors are identified through the apparent similarities between Edward Relph's (1976) and David Canter's (1977a) models of space/place, and the structuralist and post-structuralist meta-theoretical approach to space/place through facet theory and deconstruction respectively. Finally, the discourse on architecture space, in Chapter 6, is constructed through the approaches of Thomas Markus (1980s) and Bernard Tschumi (1980s~). In both chapters, multiple secondary actors are identified as all these theories develop to reflect on each other, on the margin. For example, both Relph and Canter are secondary actors in the discourse constructed on architecture space. The design of these multiple voices will be elaborated further in each space/chapter.

PRIMARY READING OF URBAN SPACE

Primary interpretation is considered as the first layer of interaction with the constructed discourse of urban space, both theoretical and empirical. However, it should be noted that data construction is itself a layer of interpretation. The primary interpretation process is based on the developed abstract framework[1] as developed in Chapter 1. This abstract framework is used to approach the underlying meaning using both the vignette of Cairo space and the theories of urban space. This primary act of interpretation represents a reflexive instance, where this framework is both the framework of interpretation and the main subject to be developed into reading strategies of urban space through this interpretation process. Accordingly, it is necessary to explore this primary interpretative framework as well as its incorporation in the interpretation process.

The primary interpretative framework of place involves people, space/place, and people-place relations and questions meaning, event, and the relations between them (R), as well as explores the urban context in relation to this abstract reading; see Figure TP.3. However, meaning and event, as will be demonstrated, are rather controversial, especially when considering the difference between Markus and Tschumi's approach. Accordingly, meaning and event, which help to question the elements, are placed at the peripheries of the primary framework. Do people-place, meaning and event signify one element, or several elements not included in the primary framework? Ultimately, this questions the internal structure of the framework and the relations between the elements of place. Are these relations represented through conjunction and/or disjunction?

▶ **FIGURE TP.3**
Abstract framework
of urban space

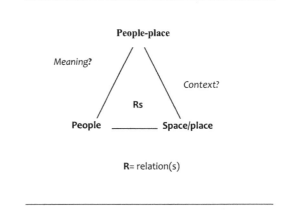

R= relation(s)

This primary framework evidently reflects the traditional paradigm of space/place which includes well-defined elements and relations. The main aim is thus to develop this primary abstract framework into reflexive reading strategies of urban space that encompass the new paradigm of urban space, with its network of complex relations between not-well-defined ~~elements~~ constituents. Accordingly, this framework approaches the constructed discourse of urban space in order to explore and reflect on itself. Elements and relations are questioned and examined in the various models to develop a better understanding of their proposed definition and nature. This framework is also used to approach the empirical discourse, the vignette of The Cultural Park for Children; however, this approach reflects on and examines both the framework and the vignette.

> [D]iscourses are not one and for all subservient to power or raised up against it, any more than silences are . . . a point of resistant and a starting point for an opposing strategy.
>
> (Foucault, 1981: 100–101)

Furthermore, this framework was developed through the generalisation of the similarities between the various models of place. This helps to provide a basic structure with which to investigate the 'silent points' in the constructed discourse of place. These silent points aid recognition of the missing parts in the discourse, the blind spot, as well as identification of any 'inconsistencies and variations' between the different constructed discourses (Alvesson and Skoldberg, 2000). Overall, this helps to enrich the interpretation with multiple theories and approaches.

The Cultural Park for Children

The primary framework is thus adopted in order to construct the primary discourse relations between the involved authors and actors. The aim here is to understand the reading and interpretation of the actors involved in the park and their consequent interventions. However, it should be noted that these questions were not used either to approach or guide the contribution of these actors. I was more interested in unstructured approaches to these actors. The data collection was therefore based on unstructured interviews and reports. Data analysis was based on the framework. Finally, the development of the park representation is discussed through the readings, perception, and conception, of the various involved actors.

Accordingly, the main question should explore the different readings of place as place, people, people-place, meaning, and event as well as the relations between. How are these readings represented through the park design? These questions are detailed as follows:

Space/place

What is the perception of space/place on both levels; context and project?
What is the interpretation of space/place?

People

What is the perception of people on both levels; society and user?
What is the interpretation of people?

Event/meaning

If approached, what is the event/meaning?

Relation

What is the perception and interpretation of the people, place, event/meaning relations?

Representation

How are these readings and conceptions represented in the architecture of the park?

On the margin of these questions and interpretative report, the discourse of the park is used to reflect the framework of interpretation, place constituents, and relations, as well as attempting to reconsider the internal structure of these constituents and relations.

This last question considers the project title used by architect and manager, re-setting (boundaries and edges), design (aesthetics, function, and structure).

Socio-architectural theories of urban space

The primary framework of interpretation was developed through a reading of both social and architecture space. It is used here to approach the primary interpretation of the social and architectural discourse on urban space as well as self-reflection of the framework through Khōra. Accordingly, the aim here is to re-approach the development of the framework rather than the exploration of an external discourse as the primary interpretation of The Cultural Park for Children.

Accordingly, the questions developed in this section approach the understanding of the different models or approaches to urban space: the model statement, its constituents – involved elements or facets – and the relations between them as well as the internal structure of the model reading.

On the margin, this reading is complemented by two reflexive reading instances, *an internal reflection* of The Cultural Park for Children and *an external reflection* through Khōra and Cairo space, re-questioning the reading of urban space.

Model statement

What is the general background and perspective of space/place?
Who is the assumed model reader(s)/author(s)?
The model statement and diagram as stated or induced by author.

Model constituents: *elements/facets*

Identify and define the proposed elements of the model.

Relations

What is the internal structure of the model?
What are the relations between the various constituents?

DECONSTRUCTION READING OF URBAN SPACE

Not taking anything for granted might be a useful rule of thumb in deconstruction.

(McQuillan, 2001: 40)

The third level of interpretation in reflexive reading is critical interpretation. According to Alvesson and Skoldberg (2000) this level of interpretation is not part of data construction. Its role is to question, 'guide and frame' data construction and the dominant interpretation which potentially allows for a new and different interpretation. In this reading, deconstruction is introduced to re-approach the primary interpretations. Deconstruction is therefore introduced as a reflexive meta-theory that forms a critical interpretation of the discourse(s) of urban space. In this sense, the role of deconstruction is to question the dominant patterns of interpretation, exposing unquestioned assumptions, inconsistencies, and internal conflicts within the developed discourse(s) of urban space and hence developing alternative interpretations (Alvesson and Skoldberg, 2000; Royle, 2000). Strategies of deconstruction are thus used as the critical level of interpretation to problematise the primary interpretation. This level of critical interpretation is applied at two different sections: the development of the theoretical frame of reference of urban space and in reflection to the vignette of 'The Cultural Park for Children' through these theories. Simultaneously, deconstruction is indirectly involved in the empirical data through the Cairo-Khōra, Tschumi's model, and social space. It should be noted that some of these themes are approached reflexively in order to develop the framework of urban space, as will be elaborated in the following chapters. This section mainly discusses the role of deconstruction as a meta-theory at the critical interpretation level, drawing on the preliminary reading of deconstruction in Chapter 2.

Deconstruction is . . . not what you think it is.

(Collins et al., 2005)

As discussed in the preliminary reading, a de-con-struction approach turns 'things upside down', denying the proposed demonstration/interpretation of data and bringing the hidden/marginal to be the centre; subverting hierarchy. Accordingly, it deconstructs this hierarchy through the simultaneous presence of two conflicting ideas, thus creating potential for alternative interpretations (Alvesson and Skoldberg, 2000). Deconstruction literally, turns things – a theory, an interpretation, an object – upside down or the other way around by searching for a conflict in representation. A visual metaphor of deconstruction using a flipping coin is presented in Figure TP.4.

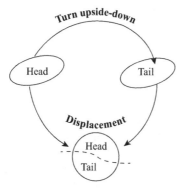

◀ **FIGURE TP.4**
The coin as a metaphor for deconstruction

Flip the coin and displace the face in search for conflict

A coin has two faces, a head and a tail. The head usually carries an image, a national representation, e.g. the queen's head, and the tail carries the coin value, e.g. £1. In this figure, a coin with head up is turned over, so the tail is face-up. Simultaneously, head and tail are displaced in a search for a conflict and potential new representation of a coin.

> Deconstruction. . . . In French, it has both a grammatical and mechanical meaning. It means both to disarrange the construction of words in a sentence and to disassemble a machine and transport it elsewhere. It also, forms a reflexive verb [se deconstuire] meaning to lose one's own construction.
>
> (McQuillan, 2001: 1)

De-, re-, construction reading space

> Badiou: when everything appears similar, nothing really is. . . .
> Deleuze: when nothing appears similar, everything really is.
> (Mullarkey, 2006: 187)

Deconstruction-reading strategies are totally dependent on the discourse being approached. They identify the structure, the inconsistencies, and the weak and missing points within the discourse. Simultaneously, they break, turn over, and change the hierarchy between centre and peripheries, and blur the edges and boundaries of the discourse. Hence, the discourse is exposed and deconstructed from inside. That is, deconstruction follows 'a logic of destabilization' that works from inside the discourse. Finally, the reading of deconstruction strategies has shown that they are related and interdependent through a singular temporal reading of the discourse (Abdelwahab and Serag, 2016).

Simultaneously, the different readings of space/place, both theoretical and empirical, were approached in the preliminary in order to identify the similarities and develop a general reading of the primary framework of urban space. However, the re-reading of these reading spaces in Part II, approaches both the destabilisation of the previous reading as well as the attempt to approach a new reading of urban space. Accordingly, the re-approach tends to oscillate between Badiou's and Deleuze's readings of similarities and differences, to identify the similarities of the differences, and destabilise the similarities through deconstruction and the notion of difference; this reading of these approaches to urban space through reflexive instances helps to develop a new approach. This reading event reflects back on the primary framework of urban space that operates on two levels towards a reflexive reading between multiple projections of urban space, one which recognises the multiplicities and dynamics of urban space beyond dualities and levels. Simultaneously, this reading event echoes the reflexive approach. Once again this follows two different paths: it acts as an external level of interpretation when approaching the constructed empirical discourse whereas it acts internally in between the models of urban space on the margin of the theoretical discourse. Subsequently, by reflecting on the deconstruction strategies as a reading event, a number of themes can be identified, which are working in the background of these strategies and which help to relate them together. These themes are represented in three sets.

The first set constitutes:

*   a rejection of a transcendental truth which exists outside the discourse
*   reading the discourse as a reality, a representation of reality, and beyond this reality
*   a rejection of the obvious or the expected reading (the traditional) which denies the potential differential reading, the unexpected
*   deconstruction is already inscribed inside the discourse
*   deconstruction is ascribed to 'the experience of the impossible'; the impossible is that which opposes the expected potential, rather than the possible. Deconstruction is impossible in a sense; it is yet to happen unexpectedly, differing from the tradition.

The second set considers the (construction) of the discourse:

*   the discourse is inscribed in the context, and vice versa through the concept of 'trace', every term holds a trace of the reality, its representation, and what is beyond
*   the discourse is constructed through sets of binary oppositions
*   centre/margin of the discourse which highlights the centre and excludes the margin
*   binary representations are included in the discourse
*   the discourse revolves around centre/margin; while metaphysics considers the centre and its de-centring, deconstruction is more interested in the margin
*   binary opposition representations follow the discourse, reality, representation, and beyond reality; metaphysics is concerned with the representation of the reality, 'facts of life'; deconstruction is concerned with binary opposition which metaphysics missed, beyond the reality representation and which is disguised through binary opposition.

The third set considers the de-construction of the discourse, embracing the margin, the cornerstone, différance, and the trace:

*   the centre/margin hierarchy is deconstructed
*   the centre is displaced through the supplement; it escapes the discourse and simultaneously is inscribed inside the discourse to demonstrate the impossibility of its construction
*   the margin is displaced through the cornerstone; the defective stone on the margin of the discourse which is responsible for the construction and deconstruction of the discourse

- deconstructing the discourse binary oppositions through différance and trace
- différance promotes ambiguity over the well-defined; it implies a continuous change of the binary terms through time and space; once you define a term it changes
- the trace blurs the boundaries between the binary opposition sets; each a term holds a trace of the other, as well as all the other terms it is not.

My deconstruction can only happen once because it is unique to the singular moment of affirmation, which is the event of my reading. . . . So, my reading (or deconstruction) is nothing more than a matter of placing myself within the operation of the text and being part of that operation (the text's own self-deconstruction) for the singular duration of my reading.

(McQuillan, 2001: 26–27)

CAIRO-KHŌRA AND SELF-REFLECTION(S)

The role of the level of self-reflection is to self-question the process, to question the author's selectivity of the various theories and vignettes of Cairo as well as dominance of the post-structuralist/deconstruction interpretation, and accordingly consider other 'representations and interpretations'. In the context of this book, this self-reflection is conducted on two levels, both internal and external. On the internal level, the book approaches two main events. The first considers the study of the empirical material through the vignette in Cairo. The second considers the study of theories of space/place. Both events are approached through the primary framework of interpretation and deconstruction critical reading strategies. An internal self-reflection on the development of the reading strategies of urban space through both these events is conducted through a reflection on Khōra, a place in between architecture and deconstruction. A simultaneous level of self-reflection reflects on Cairo space, the referential vignette. On the external level, these two events are interrupted and re-questioned. This interruption questions the book setting, both the socio-culture in between eastern and western binaries, and the academic approach through post-structuralism and social studies of place, and the empirical study of Cairo.

This section thus looks back on the book journey and considers both the questions posed and the paths taken to approach them (see Figure TP.5). It is also interested in the un-posed questions and the paths not taken. These reflections become a self critique of the selectivity and dominance of interpretative approaches and representations of the research subject, and question the potentials of taking different paths, continually developing new instances and new questions.

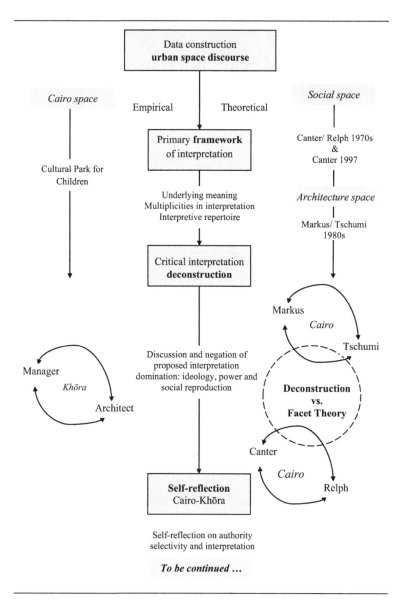

◀ **FIGURE TP.5**
Reflexive levels
of interpretation
adapted from
Alvesson and
Skoldberg
(2000: 255)

NOTE

1 It is important to note that the use of a 'framework of interpretation' in this research differs from Yanow et al. (2006).

Part II

THREE REFLEXIVE
READINGS

Chapter 4

THE CULTURAL PARK
FOR CHILDREN

This chapter presents the first reflexive reading space through a study of 'The Cultural Park for Children', at Al-Sayyida Zeinab, Cairo. The vignette 'The Cultural Park for Children' is constructed between a national and international consensus on its representation, as the park won both national and international awards, and a conflict between the park architect and designer which reflects conflicting ideas on the park design in relation to its role towards the community.

The question raised here is how the study of this vignette could help the development of the primary framework of urban space. This reflexive reading therefore is expected to play the role of both the subject of the study – the vignette in Cairo space – and the object of the study – the development of the framework of urban space. Controversially, it is hard, or rather an impossible task, to approach the study of Cairo without being overwhelmed by the empirical study of her space, as realised in Chapter 3. Accordingly, the reflexive reading of the empirical data in this chapter is approached in such a way as to reflect on both the framework of interpretation and the vignette itself. This chapter analyses a context which both traces and is traced by the space/text content – 'The Cultural Park for Children'.

Consequently, we identify the main authors, who claim authority and knowledge of the discourse and are thus involved in a conversation and/ or conflict about the representation of place. In this context, the architect, urban designer, or planner should, naturally, claim professional knowledge to urban space; whereas the manager or administrator claims lawful/political authoritative knowledge of the park. However, the manager role overlaps with designer through her interventions and contribution to the addition of a multipurpose building to the park. It should also be taken into consideration that the differentiation between the architect, urban designer, and planner is not explicit in the Egyptian context. In fact, they are all part of the architecture syndicate, despite the presence of a planning school (Faculty of Urban and Regional Planning – Cairo University). Also, the designer's claim to authoritative professional knowledge is questioned, particularly when considering public space. This role is overridden by other government administrators; for example, the role of the 'Administration of Cairo Transportation' precedes that of the architect when approaching square design, as discussed in Chapter 3. Sometimes the architect's role is taken over by government administrators and district chiefs, who come from non-architectural and non-planning backgrounds (refer to Figure 4.1).

At the same time, the narrator role is played by both the media/press and competition award committees. The press claims the authority of national representation of urban space discourse while competition committees claim authoritative professional knowledge to evaluate and award urban space. However, in this setting, both the press and the awarding committees are considered as 'commentator' rather than narrator as they both represent the discourse and help in its construction, through their authoritative claim of knowledge.

▶ **FIGURE 4.1**
A conversation construction between the press, designer, and manager based on an interview between A.I. Abdelhalim the designer and F. Hassan, the press (Hassan, 1997); an interview with the manager held by the researcher; and the Aga Khan *Technical Review Summary* of project (Akbar, 1992)

Press: Initially, everything worked very well in your plan. By the end of 1990, the garden was finished...
Ministry of Culture:

Designer: I had great hopes for this project. I wanted the garden to make a difference in the lives of the children of this community

                                    Manager: This Park is inappropriate for children...

Aga Khan: The Cultural Park for Children is a complex project if it is viewed as a product, and a significant project if it is understood as a process
Press: Now, however, visitors to the garden are told by the care takers that within the walls lies only tombs

Designer: The garden became famous as a showcase of modern Islamic architecture to be shown to important visitors

                                    Manager: The proportions of the elements of Islamic architecture used, resulted in a distorted image

Press: The garden designed originally for the children, remains void of life. Why was the garden created and simply closed?

Designer: The cultural garden was built for children, but they are not allowed to use it. It is like an expensive toy parents buy for the children but then do not allow them to play with the toy because it might break. The official reason is that there are not enough people to run the place properly.

                                    Manager: The architect wanted to build a new monument.

Aga Khan: The significance of this project lies in the evolution of the design process and the interaction with local communities as well as the architecture of the structures

Designer: In retrospect, I think it is also because we were trying to wrestle from the people their authentic culture and adapt it to the expectations of the educated...

Accordingly, the constructed discourse is approached through the primary framework of urban space in order to develop a primary analysis and report of the park discourse itself. At the same time, it is necessary to re-emphasise that the framework of interpretation helped the data analysis and interpretation rather than the construction of the discourse. On the margin, I reflect on the theoretical consideration of the discourse, theories of space/place, deconstruction, and Khōra-Cairo. The discourse of the park is hence de-, re-, con-structed through the primary framework as well as providing critical interpretation. This is followed by a discussion on the implications and development of this reflexive instant and management of empirical data for the discourse.

THE CULTURAL PARK FOR CHILDREN

In response to the rehabilitation project of the old district of Al-Sayyida Zeinab initiated by Cairo Government, the Ministry of Culture held a competition to develop the slum area of Al-Houd Al-Marsoud, a degenerated park, into a 'Cultural Park for Children'. Abdelhalim Ibrahim Abdelhalim won the competition in 1983. The park construction started in 1987 and was opened to the public by 1990 (Abdelwahab, 2009; Broeck et al., 2013). Two years later the park was granted an international Award for Architecture by the Aga Khan, which emphasised the design process involving community participation. The Cultural Park for Children involved the design and development of Al-Houd Al-Marsoud Park and Abu Al-Dahab Street, the northern boundary of the park (Figure 4.2). Significantly, the park won both a national award for the schematic design as well as an international award for the design process. This reflects a likely agreement between the national and international perspective of Cairo's space and her architectural image.

Ibn Tulun Mosque

Source: M. Abdelwahab

Al-Houd Al-Marsoud
Neighbourhood
Al-Sayyida Zeinab

Abu Al-Dahab Street

◀ **FIGURE 4.2**
The Culture Park for Children – Ibn Tulun Mosque – Al-Sayyida Zeinab

Source: IAA0269 © Aga Khan Trust for Culture/Barry Iverson (photographer); map © Courtesy of architect; Aga Khan Trust for Culture- Aga Khan Award for Architecture

▶ **FIGURE 4.3**
Key record of 'The Cultural Park for Children'

Sources: IAA0775
© Aga Khan Trust for Culture/Barry Iverson (photographer); IAA19112 © Aga Khan Trust for Culture/Barry Iverson (photographer)

Variant Name(s):
Cultural Park for Children – Formal name
Cultural Garden in Sayyida Zeinab – The press
A Community Park in Cairo – The designer
The Child's Park – The park manager
Al-Houd Al-Marsoud park – variant
Project: A competition to regenerate Al-Houd Al-Marsoud slum area as part of the regeneration project of Al-Sayyida Zeinab district initiated by the government
Designer: Abdelhalim Ibrahim Abdelhalim
Funder: The Ministry of Culture – National Project
Client: Cairo Government
Location: Al-Houd Al-Marsoud, bounded by Qadry Street, Abou El Dahab Street
Al-Sayyida Zeinab district, South region Cairo
Area: 12500 square meters including Abou Al-Dahab Street
Date: 1983-1990
Original Programme:
Entrance plazas
Open exhibition and festival plazas
The library and the media centre
Palm-tree boulevard
Green Terraces and Platforms
The museum (not built)
The theatre (not built as designed)
Nursery and the child-care centre: un-built and replaced by the library
Multi-purpose unit (added on manager request)
Abu Al-Dahab street (Harah):
includes a café, fountain, community room, prayer space, five book and craft shops, festival and community plazas, trees, and steps
Award: The Aga Khan Award for Architecture in 1992

Above: The main building

Above: Haret Abu Al-Dahab

Today, a significant change in the park is evident. The realities of the park's everyday life, and Abu Al-Dahab Street in particular, show indifference to the original design scheme, creating conflicting community activity and isolating the park from its context, Al-Sayyida Zeinab. This sparks interest in the park discourse, what the park is trying to tell us about urban space through her story.

It is also interesting to reflect on the variant names of the park as they reveal the different readings and understanding of the park. The official name of the project is 'The Cultural Park for Children'; the park designer mostly refers to 'community park'; while the manager simply describes it as 'the child's park'. In between the 'community' and 'child' lies a conflict between the architect and the manager about the role of the park and the appropriateness of the park for its function.

In order to understand the story that the park is telling through these changes and conflicts, I will attempt to reconstruct the conversation

◀ **FIGURE 4.4**
The park schematic
design – national
award 1987

Source: IAA19112
© Aga Khan Trust for
Culture/Barry Iverson
(photographer); http://
archnet.org/library/
files/

◀ **FIGURE 4.5**
The opening of
the park

Source: IAA19109 ©
Aga Khan Trust for
Culture/Barry Iverson
(photographer); http://
archnet.org/library/
files/

► **FIGURE 4.6**
Changes to
accommodate
children activities
2008

between the actors involved in the park, and explore the relation between the different readings of urban design through design, management, writing, and prize awarding; see Figure 4.8. The discourse of the park is thus approached through the construction of a conversation between the different actors involved in agreements – between the national and international awarding committees – and disagreements – between architect and manager. The constructed discourse is particularly interested in the dialogue, or rather the conflict that was developed between the two main actors, architect and manager, around the park design, reading, and interpretation in relation to Al-Sayyida Zeinab, the community and the child. However, this is not an attempt to criticise or defend either the architect or the manager or to explore the success and failure of urban space. However, the argument between architect and manager was crucial in building up the park discourse in order to explore the relation between the different readings, and the approach to the park interventions. It should also be noted that the user is not directly involved in this discourse. The actors' reading and representation of the user is considered more important in this context.

At the same time, the press and the Aga Khan play the commentator role. The discourse is therefore built up through the architects, published interviews with the press, and academic writings; telephone non-structured interviews with the manager; journal and magazine articles and interviews; and the Aga Khan 'Technical Review Summary' of the project. This is complemented by archival photos of the park on its opening in 1990 by the architect and the Aga Khan; and recent photos I took in 2008, eighteen years later. I am also involved in the discourse through the interviews, the visits, photos, observations and content analysis; see Figure 4.7.

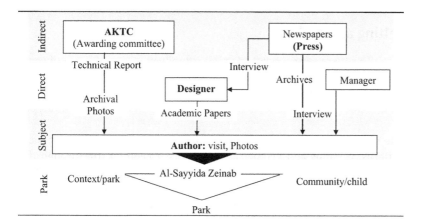

◀ **FIGURE 4.7**
Data collection and building the discourse for The Cultural Park for Children

Source: Aga Khan Trust for Culture

Figure 4.8 is an attempt to represent the de-, re-, construction of the park discourse through the identified actors. The reporting of this interpretative repertoire thus involves a four part-cycle: place setting, reading, conception, and representation. The first part attempts to describe the urban setting. It starts with a basic record of the park and a general introduction to the project; this is complemented by observations about the urban space specialty. The second and third parts are developed through the construction of a conversation between the involved actors, the architect, the manager, as well as the Aga Khan international award committee and national press. As discussed, this part considers the development of the park through the different readings of these actors as viewed through the primary framework

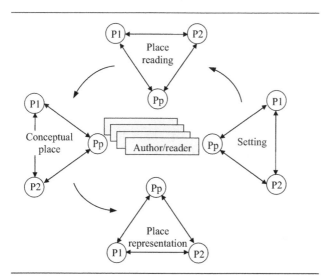

◀ **FIGURE 4.8**
Primary interpretative report of the park – P1= Place, P2= People, Pp= People-place

of urban space. The last part considers the representation of the park by the different actors. However, the cycle is not closed, as the park/urban space setting is developed through the multiple readings of all the involved actors, including the user for example who are not included in this study.

Setting and authors

> This theme was developed architecturally in varying degrees of complexities, from the simple spiral walls of the entrance, to the complex cascades of vaults and domes forming the spiral museum.
>
> (Abdelhalim, 1986: 68)

The park design contained two layers: the transformation of the physical patterns, rhythms and geometric patterns (see Figure 4.10), through the elements defined within the container-place; and the ceremonial building process that helped to re-approach the spatial design. Thus, the architect develops the park design through approaching the following objectives:

- the park plan and organisation draws on the Ibn Tulun spiral minaret
- the existing palm trees formed the co-ordinates of the geometric patterns
- the park rhythms and stylistic feature draws on the remains of the Islamic architecture style in the neighbourhood.

▶ **FIGURE 4.9**
The park architectural plan

Source: IAA0267 © Aga Khan Trust for Culture/Barry Iverson (photographer); http://archnet.org/library/files/

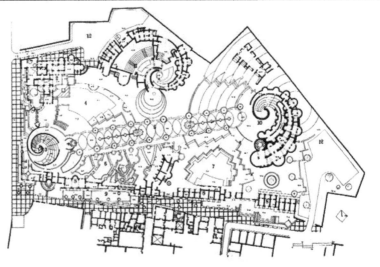

		9 Abu Al-Dahab street development
1 main entrance	6 terraces	10 children museum
2 library and atelier	7 Abu Al-Dahab entrance yard, theatre	11 theatres
3 main fountains		12 service entrance and yards
4 green yard	8 Abu Al-Dahab community wall	
5 palm tree promenade		

◀ **FIGURE 4.10**
In-context: rhythm
and geometry
pattern

Source: IAA0267
© Aga Khan Trust for
Culture/Barry Iverson
(photographer)

I really never thought of [The Cultural Park for
Children] as Islamic nor of paradise [Islamic
garden image of heaven] as I was designing. . . . I
began to realise that, they are not apart from the
concept of paradise.
(Abdelhalim, 1986: 68)

Kant described space as neither
matter nor the set of objective
relations between things but an
ideal internal structure, an a priori
consciousness, an instrument of
knowledge.
(Tschumi, 1975a: 29)

The park designer summarises his philosophical
orientation as:

- the rhythms, ruling the symbolic relations of place
- the formal geometric patterns
- the co-ordinates of these patterns
- the community rituals, the building ceremonies
  which is an important aspect to the designer.
  (Akbar, 1992; Abdelhalim, 1983)

These themes arise from the architect's perspective
of both the project in context and his role as an
architect. The park is considerably developed through
a projection of the architect's theoretical background
rather than through an objective reading of urban
space, which he describes as 'an actual community
building project in Egypt – a cultural park for
children near Ibn Tulun Mosque in Cairo – [which]

gave me the opportunity to test out my theories'[1] (Abdelhalim, 1988: 140). This reading is braced by Abdelhalim's belief in the park as 'an educational instrument' (Abdelhalim in Saleh, 1989) and the ability to 'change society through architecture' (Abdelhalim, 1996: 53; Abdelhalim, 1989).

This chapter thus explores the architect's reading and interpretation of urban space. The developed argument is that it is his potential reading of the community, rather than of the child, and the urban context as old and historic rather than a new conception, has influenced the architect's design and representation of the park.

Container space

a setting 'in which bodies [/matter] are located and move' (McDonough, 2014).

It is important to highlight the architect's approach to space/place as a container space, topos. 'The configuration of a procession and the structure of poetic rhythm reflect belief and ideology in their relationships in space. The task of the designer is to disentangle these containers of order and discover their underlying geometry' (Abdelhalim, 1988: 143). A container space, as discussed in Chapter 1, 'dominated senses and bodies by containing them' (Tschumi, 1975a: 29). In this sense, though the urban space is the container of both people and event(s), the relation between place, and people, and event(s) is minimised so that urban space can exist alone; like the monuments frozen in time and space.

Therefore, Abdelhalim's reading of Al-Sayyida Zeinab and conceptualisation of the park as well as the relation between the architect – himself – and the community developed his philosophical perspective and approach to the project. The park design contains two layers, the physical and the social; the transformation of the physical patterns; and the ceremonial building process.

On the other hand, the manager holds a different approach to the park and Al-Sayyida Zeinab district. Fatma Al-Ma'doul is an ex-resident of Al-Houd Al-Marsoud neighbourhood. Her old house, where she spent her childhood, lies behind the park. She remembers the old park of Al-Houd Al-Marsoud, built in the late nineteenth century: 'a large beautiful garden that contained a collection of rare plants'. She also recites the deterioration of the park through the years, since the war in 1967 when air-raid shelters were built in the park. The deterioration process of the old park continued over the following years as several plants were destroyed, administrative buildings added, and many workshops occupied the place.

However, Fatma still remembers the beautiful image of the old park of Al-Houd Al-Marsoud.

◀ **FIGURE 4.11**
The deterioration in Abu Al-Dahab Street

She also has an educational background in applied arts; she is especially interested in Islamic art and architecture which reflects the traditional image of Al-Sayyida Zeinab. The manager hence holds a romantic and nostalgic perspective towards the park and community, which she projects on from the wonderful past. She considers the physical setting of the park as part of her identity and childhood experience as an ex-resident. Accordingly, she appreciated the contribution made by the park in turning a slum area into a beautiful garden, and removing many illegal activities. However, she accuses the architect of misunderstanding both the true nature of the district and the child.

It is also important to explore the background of the Aga Khan Institute which awarded the park an international Award for Architecture in 1992. The Award for Architecture was established in 1977, (AKAA). It is a part of the Aga Khan Trust for Culture, whose objective 'focuses on the physical, social, cultural and economic revitalisation of communities in the Muslim world' (Aga Khan Trust for Culture). The Aga Khan is the Imam (spiritual leader) of the Shiaa Ismaili Muslims, descendant of Fatima, the daughter of the Prophet. Also, the Fatimid Empire, the founders of Fatimid Cairo 969 AD, (Beeson, 1969), is called after Fatima. The Aga Khan interest in Cairo is evident through the University of Al-Azhar, The Academy of Science, Al-Azhar Park, and 'indeed the city of Cairo' herself, (Ismaili-Studies, 2007). The Aga Khan approach to Cairo is hence deeply embedded in the image of the old Islamic city founded by the Fatimids, their heritage.

▶ **FIGURE 4.12**
'The park turned a
slum-deteriorated
area to a beautiful
garden' (based
on the manager's
comment)

Place

As the architect walks through the urban space,
he sees the minaret of Ibn Tulun mosque, 879 AD
standing in between the old, dense, and poorly
maintained district of Al-Sayyida Zeinab. '[O]ne
of the oldest, most densely populated and poorly
maintained quarters in Cairo' (Abdelhalim, 1988:
140). He perceives the city as poor and decaying,
and the monuments standing frozen in time; i.e. the
past is creative whereas the present is a remnant
of that past (Abdelhalim, 1989; Abdelhalim, 1988).
Simultaneously, The Cultural Park for Children
is built on the remnants of Al-Houd Al-Marsoud
park. Abdelhalim explores the existent rhythms,
geometric patterns, and their co-ordinates within
the old neighbourhood; and identifies the objects
within the place as the landmarks, the minaret
Ibn Tulun, the domes of mausoleums, the land
patterns, existing palm trees from the old park, and
so on. These objects within the container are hence
analysed and synthesised to develop the park theme
and design. The architect's place is conceived as a
spatial container of Islamic historic references that
reads and interprets the spatial patterns within both
the park setting and Al-Sayyida Zeinab district.
Thus, the architect's role is hence an interpretative

synthesis of these spatial features within the new place which in turn forms a new layer juxtaposed to the old one.

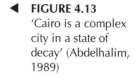

◀ **FIGURE 4.13**
'Cairo is a complex city in a state of decay' (Abdelhalim, 1989)

◀ **FIGURE 4.14**
Ibn Tulun Mosque. 'Monuments stand as a witness embodying a moment of creativity.' (Abdelhalim, 1989: 237)

Interestingly, the Aga Khan holds a different reading and interpretation of the same old district of Al-Sayyida Zeinab. The historic district is perceived as a rich place 'composed of different urban configurations and offers several layers of architectural expression, . . . a rich variety of monuments . . . residential buildings from the ottoman and late Mamluk period . . . side by side modern and more recent buildings' (Akbar, 1992: 1). In other words, the Aga Khan, as an international institute, holds a different perspective of Al-Sayyida Zeinab, and Cairo in general, from the architect. The Aga Khan considered the old city to be a rich place with considerable potential and reflecting several layers of history as against the architect's monolithic reading of the old Islamic city as standing and struggling within a contemporary poor and decaying context. Controversially, the Aga Khan background shaped the emphasised interest in the image of the Islamic period.

The manager's place is manifested through her intervention and place adaptation, as well as her criticism of the architect's approach. Her interventions into the park are primarily related to the place function and will be discussed in the following section. She reflects on the park architectural expression and aesthetics. On the one hand, she questions the use of the monolithic image of Islamic architecture to represent a multi-cultural place like Al-Sayyida Zeinab. On the other, she disapproves of the use of a distorted image of features of Islamic architecture; the architect's synthesis involved the manipulation of the proportions and geometry of these features.

It is necessary here to reflect on the Aga Khan report of the park design as one which 'respects and integrates with the site' through the synthesis of 'traditional forms with modern needs' (Akbar, 1992: 6).

▶ **FIGURE 4.15**
The synthesis of the elements and features of Islamic architecture

Source: IAA1068
© Aga Khan Trust for Culture / Barry Iverson (photographer)

People

People are a crucial element in the discourse of the park. The 'people' relation to the park gained the attention of the Aga Khan Award; it is also deeply embedded in the conflict between the architect and the manager. It has been exceptionally hard to draw on the people discourse unrelated to the other elements, place, people-place, and event. The people discourse was developed through the exploration of the architectural programme, which is considered as the architectural manifestation of people, as well as people's accounts by the architect and the manager.

> [T]he insertion of the park into this congested urban fabric has gone far beyond the original brief. It has generated a renewed sense of community by extending its presence into the surrounding streets. The residents take pride in their neighbourhood as well as their park.
>
> (The award jury note: Aga Khan Trust for Culture)

The park programme was developed through a dialogue between the architect and the community and continued through the manager interventions in the park. This reading of the people involved the child, the expected user of the park (inside) as well as Al-Houd Al-Marsoud residents, and the neighbourhood community (outside the park). The different readings of the park considered the child or the community at the centre of the discourse or else at the periphery (Figure 4.16).

People/park reading

Child
User experience
Inside

Socio-culture context
Al-Houd Al-Marsoud
community

Outside

◀ **FIGURE 4.16**
Reading people in
the park

◀ **FIGURE 4.17**
'People' discourse
oscillates between
the children and the
community

Source: IAA1068 ©
Aga Khan Trust for
Culture/Barry Iverson
(photographer);
IAA19131 © Aga Khan
Trust for Culture/Barry
Iverson (photographer);
http://archnet.org/
library/files/

'All users, neighbours and officials love the project and are proud of it' (Akbar, 1992)

The Aga Khan Institute recognises the users of the park as children in elementary and intermediate school; i.e. their ages range between 6–15 years. However, it also recognises and emphasises the relation between the park and the residents of Al-Houd Al-Marsoud neighbourhood, the community. Accordingly, it highlights the community participation that started through the Cornerstone Ceremony, the event described by the architect as the building ceremony; and manifested through the development of Abu Al-Dahab Street.

The 'building ceremony' and 'Abu Al-Dahab Street' will be both discussed in the following sections. As already explained this section is more interested in the architectural programme development.

The architect's reading reveals his interest in the design process and its manipulation but takes little account of the child playful experience in place. He considers the community within the wider socio-cultural context of the park rather than the child's experience inside the park, i.e. the child was marginalised in the architect's approach.

> Walls are not high and thus any [child] fall from
> the [high] wall will not cause injuries. Thus far,
> we have had no such incident.
> (Abdelhalim in Akbar, 1992: 6)

▶ **FIGURE 4.18**
Children playing on the roof

Source: IAA19127 © Aga Khan Trust for Culture/Barry Iverson (photographer)

The architect described the park as a 'toy' and one that was too expensive for the socio-cultural background of the child. Interestingly, it seems like the architect was actually attempting to build a toy, or rather an 'educational' toy, where 'teachers use the variety of forms to demonstrate the principles of geometry' (Akbar, 1992: 5–6). The park was designed to educate the child about geometry, history, and nature through a guided exploration of the park elements. However, the space provided for the child's free activities and playing was confined to a minimum, and represented by a single slide beside the rear wall of the park, as is evident in the archival photos of the park opening day.

◀ **FIGURE 4.19**
The children's designed playground confined to a small area by the rear wall

Source: IAA19130 © Aga Khan Trust for Culture/Barry Iverson (photographer)

The Aga Khan considered the community children as poor and the park as an opportunity for them, 'a chance to play in a beautifully arranged space' (Akbar, 1992: 5).

The manager, on the other hand, asserted firmly that the park was inappropriate for child use. She criticised the architect for not including the child in his design approach. The architectural programme did not provide outdoor shaded areas for child activity. The indoor spaces were also too small to accommodate different activities. The architect chose traditional stone construction for the main building; the wall bearing system helped to produce small spaces. The stairs, fountains, and other finishing materials are hard, and have sharp edges and accordingly are unsuitable for children playing and running about.

The manager's reading of the park reveals her administrative concerns, and her special interest in the child's experience in the park in comfort and safety. Accordingly, she asked for a multipurpose indoor space to accommodate the various children's activities, especially on hot and/or rainy days. The original park architect, A. I. Abdelhalim, refused to design the new building. This precipitated an argument between the architect and the manager. Finally, a multipurpose building was designed and built with the help of another architect (Figure 4.21). She continued to intervene in the

▶ **FIGURE 4.20**
Sharp edges and finishing material unsuitable for children

▶ **FIGURE 4.21**
A multipurpose building added to accommodate children in the park

▶ **FIGURE 4.22**
The park spaces are adapted to the children's activity

park to accommodate children's needs, providing shaded areas for their activities, removing most of the wooden artefacts and other finishing material that was inappropriate for child use. Accordingly, the child's place in her park is everywhere, on the flooring, along the walls. Figures 4.23 to 4.26 introduce a visual representation of the manager's interventions in the park based on photos of the park in 2008 and Aga Khan archival photos of the park from the opening in 1992.

◀ **FIGURE 4.23**
Re-adapting shaded areas for children's activities, 2008

▶ **FIGURE 4.24**
The landscape finishing material, 1992

Source: IAA19115
© Aga Khan Trust for Culture/ Barry Iverson (photographer)

◀ **FIGURE 4.25**
Removing the wooden structures to facilitate the children's movement, 2008

▶ **FIGURE 4.26**
The use of the traditional wood crafts in the park, 1992

Source: IAA19114
© Aga Khan Trust for Culture/ Barry Iverson (photographer)

Around 1968, together with many in my generation of young architects, I was concerned with need for an architecture that might change society – that could have a political or social effect. . . .

However, both through facts and through serious critical analysis [demonstrated], the difficulty of the imperative . . . all shared the skeptical view of the power of architecture to alter social and political structures.
(Tschumi, 2001: 5–15)

As discussed the architect's reading of 'people' extended beyond the user of the park, to the community. However, his reading of the community appears to have changed over time and through interaction with the community. The architect approached the community through a belief held by his profession: 'the role of the architect and his responsibility to interpret, and understand the culture and change the society through his architecture' (Abdelhalim, 1996; Abdelhalim, 1989: 236). Accordingly, he considered the community as poor and needy. However, he acknowledged that the changing dynamics of the community are never stable and are strongly related to the culture and identity of each community (Abdelhalim, 1996). And the architect's role is to re-establish the connections between this culture and the built environment. Accordingly, the architect must understand the culture of the community and learn from this rather than simply applying outside rules which he identifies as 'neat tidy plots' (Abdelhalim, 1996: 53).

Interestingly, during the building process, people got the chance to play an active role together with other commentators on the development of the project. This is quite evident through their influence on the development of the park's architectural programme. A nursery and child care centre were introduced as key elements in the programme, a service that reaches into the community to help working mothers. However, the community argued for a change in the programme to include a library which it felt would be more appreciated by the children of the community. Nurseries and child care are already sustained within the community 'on a co-operative basis' (Akbar, 1992). The community also influenced the programme for Abu Al-Dahab Street, 'to include a café, fountain. Community room, prayer space, books and craft shops, festival and community plaza, trees and steps' (Akbar, 1992). We will explore further the development of Abu Al-Dahab Street in the park re-setting section.

[A]rchitecture and its spaces do not change society, but through architecture and the understanding of its effect, we can accelerate processes of change underway . . . [or] slow down.
(Tschumi, 2001: 15)

This role also involved attracting the community into participating in the park production rather than playing a passive role. By the end of the project, the architect's approach to the community changed radically:

> we were trying to wrestle from the people their authentic culture and adapt it to the expectations of the educated. We wanted to intellectualize it to the expectations of the educated.
> (Abdelhalim in Hassan, 1997: 13)

Accordingly, three approaches could be read through the architect's reading of people: educational/professional (the architect can and will teach and change the community); educational/ interpretative (the architect learns from the community); and finally the opponent (the architect struggling with the community to change and the community fighting back against change). The architect's reading of people emphasises the value of the history of Al-Sayyida Zeinab – 'the community draws strength and pride from its reservoir of history' (Abdelhalim, 1996: 54) – which extends to the 'event' through the 'building ceremony', a reading through the community rituals and festivals, which will be discussed further through the event. And, this would need to be reflected back into the architect's reading of people through the event. The discourse of the event is therefore constructed through a reading of the architect's event, and the park as an event.

People-place: meaning

As discussed, the architect's reading of people has emphasised the reading of the community rather than the child. Interestingly, the architect's reading of the people-place relation emphasised the child's relationship to the park as the prime generator of the formal pattern of the park. The park design

> was drawn from the interpretation of the Ibn Tulun minaret. The spiral of the minaret, clearly visible from the site symbolises the idea of growth, which was taken as the main theme for the Park, to give form to what was common between children and the Park . . . life.
> (Abdelhalim, 1986: 68)

This was complemented by another layer of reading of the community rituals through the Sayyida Zeinab Festival; 'more important than these monuments, however, is the lively festival of Sayyida Zeinab held every year, during which the identity and the culture of the community is re-enacted and regenerated' (Abdelhalim, 1988: 142). The latter reading was developed through the 'building ceremony', which is discussed through the event.

Forms do not follow functions, but refer to other forms, and functions relate to symbols . . . becomes a syntax of empty signs . . . derived from a selective historicism that concentrates on moments of history (Tschumi, 2001: 11).

► **FIGURE 4.27**
The architect's
reading of people-
place relations
(meaning) in
between the child
and the community

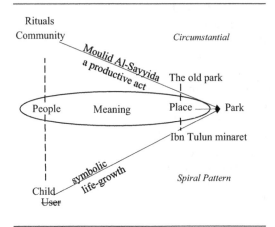

Ultimately architecture frees itself
from reality altogether . . . [where]
the excesses of meaning [style]
have emptied the language of
architecture from meaning. . . .
How can meaning be produced
when signs only refer to other
signs when they do not signify but
only substitute? (Tschumi, 2001:
37, 176).

Accordingly, Figure 4.27 presents a reading of the
architect's approach to people-place relations.
The architect appears to approach this relationship
through his reading of the meaning of place rather
than the meaning people attach to that place.
The child-park relationship is symbolic of life and
growth and found representation through the
spiral of the Ibn Tulun minaret. Consequently,
the community-park relationship is considered
through the productive act of the Moulid, Al-
Sayyida Zeinab annual festival that brings the
community back into the park. And through the
'building ceremony', the architect added another
layer, a 'circumstantial layer', to the original formal
design inspired by the minaret. In summary, the
architect's reading of the people-place relationship
considered neither the community nor the child
as users of the park. He was more interested in the
symbolic value of people.

The manager, on the other hand, emphasised
the functional meaning of the park. She
approached the child as a user. Accordingly, her
interventions in the park involved adapting it
for children's needs, providing shaded areas and
indoor activity spaces which were not included
in the architectural programme, and removing
sharp and hard finishing materials provided
by the architect and deemed unsuitable for a
children's play area.

Event

> A garden, park, or a small landscape project can be a valuable instrument to trigger and set into motion a community-wide process which can uncover, re-establish, and perhaps reseal the gulf or the rupture caused by modernity and industrialization, and reaffirm the culture of that community and the production and sustenance of its environment.
>
> (Abdelhalim, 1996: 54)

The reading of the event in this section differs from the reading of the event in Chapter 2, which considered the concept of event in metaphysics and in deconstruction. The event, in this section, takes a step away from theory to approach the concept of event through the readings of the authors of the park, the architect and the manager. However, these readings reflect on the presence of the event in space and time.

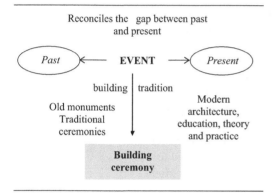

◄ **FIGURE 4.28**
The architect's planned event

The architect's reading of the park revolved around two main lines of thought. The first expressed a continuous nostalgia for the past and tradition; place is a decaying remnant of the glory of the past; and more importantly, it is scene of the continuing tradition of Al-Sayyida Zeinab festival, ceremony, and rituals. The second line of thought expressed the authoritative position of the architect towards the park through his professional knowledge; the role of the 'educated' architect in understanding and interpreting people and place in order to 'change the society' and 're-establish' new relations. The architect's reading thus reflects two images, the glory of the past through the standing monuments and continuing festival of Al-Sayyida Zeinab, and the architect's modern western education, modern theories, and practice.

Building ceremony

The architect thus introduced the event of the 'building ceremonial', which in his words 'suggests some kind of contradiction since building involves construction, finance, and law, while ceremony is associated with rituals, festivity, and regeneration' and in simple words is 'the integration between both culture and production' (Abdelhalim, 1988: 140). This building ceremony is a construction event that considers the building process rather than the architectural space in operation. Also, it is deeply embedded in the history and traditional culture of place and community, as will be explored further.

Spatial

One intention of the event of the building ceremonial was to show the construction process of the park to the community of Al-Sayyida Zeinab. This is not a part of Egyptian culture, where buildings under construction are covered in tents until opening day to hide them from the onlookers. Accordingly, a full-scale model of design, fashioned from wooden poles and canvas was hence on the event, with platforms and terraces marked on the ground. At the same time, dancers, musicians, and artists were invited to participate as well as the community of Al-Houd Al-Marsoud (Abdelhalim, 1988).

Historic

Simultaneously, the building ceremonial involved the laying of the cornerstone of the park, reflecting building rituals from the time of the pharaohs, and Islamic building traditions up to the present day in order to celebrate the foundation sacrifice.

Socio-cultural

Also, the ceremony reflected the interest in Al-Sayyida Zeinab Festival (Al-Moulid); the ceremonial processions together with the rhythm of the folklore dance and music were involved in the event. This helped the initiation of a connection between the park and the community, which gradually involved them in the design process and programme: 'a link between the activities of building and the culture of the community' (Abdelhalim, 1988: 148).

Political and economic

Finally, the building ceremony was held on the occasion of the Child's Festival in Egypt, which occurs annually in November. The president and his wife and the minister of culture attended the event which also included the laying of the project cornerstone (Abdelhalim, 1988). The involvement of this political party in the building ceremony, the image of hundreds of children playing and dancing around the full-scale model of the park . . . with thousands of citizens, attracted the presidential attendance as well as helped to re-initiate political interest in and funding for the project, which had stopped some time ago (Abdelhalim, 1988).

After the event, things settle down, intensities dissipate. The memory of the event remains: not as image or recollection, but as kind of field of virtual potential never quite exhausts itself in the process of becoming more than it never (actually) was. (McCormack, 2008: 8)

The difference between the architect's approach and deconstruction to the event is evident. Derrida, on the one hand, is interested in the event yet to come, unexpected and unplanned, and a part of successive events in the future. Abdelhalim, on the other hand, is planning an event which is both embedded in the historic culture of place as well as becoming a new instance in the history and culture of the park, to be completed and finished before the production of the space into the everyday life of the community, as a memory and a past recollection of place.

. . . and the child running about

> [The architect]: We set up a life-sized model
> of the garden on the grounds and watched the
> children move from one area to the other.
> (Hassan, 1997: 13)

> The manager on the other hand was interested in
> the production of the event through the interaction
> of the child with the park through everyday
> spontaneous activities.

◀ **FIGURE 4.29**
The Children's Park represented along the walls of the park

The controversy in reading the event lies in its definition as an attribute of the dynamic relations of place within its wider context, which implies the existence of multiple events and/or multiple existence of the event. This is evident in the reflection on the perception of place as event of both the architect and the manager. Both acknowledged the past of the place and the intrinsic value of its heritage. The architect embedded the event into this past, perceiving place as a container of monument and heritage elements, community rituals, and traditional building ceremonies. This heritage was defined in relation to the Islamic foundations of the city. However, the relations between place, people, and event(s) in everyday life is minimised such that place can exist alone like a frozen monument in place and time. On the other hand, the manager embedded the event in a place identity that considered the potential of its future development. Place was not decaying but attached to the past image of a better quality urban space that could be achieved again.

ON THE MARGIN

This chapter explored the readings of 'The Cultural Park for Children' by the architect and manager as well as other involved commentators, through place, people, people-place, meaning, and event. This reading has simultaneously emphasised the exclusion of a user's reading. On the margin, the deconstruction-reading event is reflected through the two interrelated triads of reading strategies introduced in Chapter 2: meaning: cornerstone: binaries and event: context: history. This deconstruction reading thus reflects on the park name, boundaries, and edges and design.

This is a reading of the binaries involved in the discourse of the park, which is mainly presented through the architect/manager binary as reader and author of the park. The first binary is read through the architect's container space of past monuments, rhythms, etc. whereas the manager considers a relational reading of the park: in relation to her old house, to the neighbourhood, to the community, etc. Another binary is represented as past/present, the past history and traditions of building and the present everyday activity of the child in the park. Consequently, the architect's reading has subverted the child in favour of the community as users, whereas the manager emphasised the role of the park towards the child. The architect's reading extended to highlight the symbolic relation between community rituals as well as child growth and the park design in the architect's.

Beyond these representational binaries, though on the margin of this reading, the manager is a community member, an ex-resident of Al-Houd Al-Marsoud neighbourhood, a user. Reading the park through a struggle in between architect and manager, (Hassan 1997) hides a further struggle between the architect and a community member; in this case a community member who possessed the power to approach the park as an active author holding professional authority, the manager.

Architect	/	Manager
Place		
Container	/	Relational
People		
Community	/	Child
People-place: meaning		
Symbolic	/	Identity
Event		
Past	/	Present
~~Architect~~	/	~~Manager~~
Architect	/	User

The manager intervened in the park design, making several changes to adapt the place for children's use, according to her perspective and reading of place. Consequently, the binary was reversed to subvert the architect in favour of the user, user/architect, in her reading.

Simultaneously, the reading of the park is embedded in the event of this subversion, the empowering of the user. The binary representations reflect on the context of this event; first, the architect, claiming professional knowledge and attempting to educate the child and change society through his park, his theory, and his practice; and second, the manager, having an administrative authoritative knowledge, attempts to imprint this authority on the park, to emphasise the child and society relationship. Author/reader binary are deconstructed; not only is the author the first reader of his/her work, but simultaneously the reader is also the author of the park. This subverted reading is thus projected through the name, boundaries. and design of the park.

User	/	architect
Name		
Child	/	community
People		
Community	/	Child
Edges	and	boundaries
Community	and	relational
Relational	/	container
Design		
Minimalist	/	maximalist
People-place: meaning		
Identity	/	~~symbolic~~
Space	/	programme

The name

The park name was presented as a binary between the child's and the community park, in between the manager (user) and the architect respectively. The binary representation of the park name is also projected through the architectural programme (Figure 4.3). The architect, on the one hand, proposed the inclusion of a nursery and a child care centre in his primary programme to support working mothers in the community. The community, on the other hand, asked for the inclusion of a library to serve the children of the community in place of this centre, and this was eventually built. This incident reflects the importance of the inclusion of the community in the development of the programme. It is worth noting that the intended user – the child as identified by the community – however was not included. It also emphasises the architect's reading/misreading of the park through the wider social community – a maximalist approach, rather than the actions and activities of the user (child) in the park. Consequently, the community approach deconstructs this binary representation, since the community acts as spokesman for the child.

Boundaries and edges

Another binary was presented that read the park space as between a container and a relational space. The architect's reading of the container space of the park, which holds time: history, monuments, and traditions, and socio-cultural values inside this container. However, the architect's maximalist approach to the park design aimed to extend the park space outward, to reach the community, the context, identified as inside the neighbourhood community. Accordingly, he rejected the development of well-defined boundaries to the park within the neighbourhood. The park edges are designed as permeable walls that allow access between park and surroundings. Abu Al-Dahab Street was designed to facilitate and encourage this movement. The architect's emphasis on the container space simultaneously highlights the relational spaces – between park and neighbourhood – inside the container – the neighbourhood. The manager's relational space, on the other hand, is contained within the park's solid boundaries, which is again supported by the social realities of the community. The park needs to provide a safe enclosed space for children, in the absence of parental supervision. Beyond the design scheme, Abu Al-Dahab Street continued to deteriorate, building an unhealthy barrier between the park and her surroundings (Figure 4.11). The materialisation of the concepts of permeable boundaries and edges between the inside and outside physical space turned problematic.

In this reading, the primary framework of interpretation of place has helped the de-, re-, construction of Cairo space through the case of 'The Cultural Park for Children'. In the following chapters, 5 and 6, the reading of social and architecture space thus aims to re-approach the development of this framework.

NOTE

1   The designer theory of community event is explained in his doctoral thesis (Abdelhalim, 1978).

Chapter 5

SOCIAL SPACE

This chapter reviews the second reflexive space, which consists of two parts. The first examines Canter's (1977a) and Relph's (1976) early models which reflect a positivist, well-defined reading through linear and direct relations. And the second looks at the facet theory (Canter, 1997) that attempted to respond to the criticism to Canter's earlier model that excluded architecture space, particularly the physical space, through the inclusion of Markus's discourse on architecture space. However, the facet theory of place differs from Canter's early approach to place; to put it simply the facet theory, like deconstruction, is a meta-theory rather than a theory. This brings about the inclusion of deconstruction into the built discourse of social space in order to reflect on the facet theory rather than just to explore deconstruction itself as already covered in Chapter 2. Both facet theory and deconstruction recognised the dynamics and complexities of urban space; however, the facet theory works through a pre-defined framework and system of categorisation to structure the reading of social space. At the same time, the social space discourse is theoretical, concerned with the development of theoretical approaches to place. However, these theories are intrinsically developed through empirical approaches as well as aiming at the development of empirical studies of place. Simultaneously, the reflexive reading of the discourse involves the primary interpretation through the primary framework developed in Chapter 1, as well as the de-, re-, construction of the discourse through critical interpretations and self-reflections, Cairo-Khōra.

In a similar construction to the empirical discourse on The Cultural Park for Children presented in the previous chapter, this reading attempts the construction of the social space discourse. The discourse on social space is thus constructed through the models of place developed in the late 1970s by Canter in environmental psychology and Relph in human geography; together with Canter's development of the facet theory of place in the 1980s and 1990s through the integration of structuralism, architecture, and environmental psychology. Accordingly, the authors of the discourse on social space are identified as Canter and Relph on the one hand, and facet theory and deconstruction on the other, and reflections

▶ **FIGURE 5.1**
Social space
discourse setting
through reflexive
reading between
the discourse actors

Author 1a: Canter 1977a: Place cognition Author 1b: What is facet theory?	On the margin, Cairo-Khōra Cultural Park for Children
On the margin, Cairo-Khōra Cultural Park for Children	*Author 2a: Relph 1976: sense of place* *Author 2b: What is not facet theory?* *Deconstruction*

Groat (1981) has demonstrated how Canter's approach and social studies of place in general have emphasised meaning, the signified, over the physical form, the signifier. On the other hand, 'architectural theorists' have tended to focus on the physical place rather than the meaning (Groat and Després, 1991).

through Cairo-Khōra on the margin. The different theoretical perspectives involved are used as commentators to reflect on the development of the conversation; see Figure 5.1.

EARLY MODELS OF PLACE

> My understanding of place is almost opposite to that of David Canter.
> (Relph, 1978: 237–238)

The review in Chapter 1 demonstrated how both Relph in 1976 and Canter in 1977a proposed two similar three-fold models, although they worked within different disciplines and epistemological backgrounds. Consequently, several studies have either directly or indirectly referenced both models, though especially Canter's (Groat, 1995a; Groat and Després, 1991; Sime, 1985). Sime (1985) therefore recommended 'a combination of the two and incorporating more architecturally based arguments' (Sime, 1985: 34). This recommendation is adopted through this reflexive reading of both models. The similarities between the models are evident. Groat (1995b) accordingly considers the opportunities for integrating the phenomenological and empirical approaches in social studies through the study of these 'similar' models. However, these apparent similarities discussed by many authors, are considered as contradictions and oppositions by the authors of both models (Canter, 1977b; Relph, 1978). Accordingly, these similarities and oppositions are explored in the following sections.

Social space projection

Cognitive theory

emphasises the rational Mind, the conceptual Space, relationship to the Body (social behaviour) and Space (experience)

Canter's model was developed through a 'psychologist approach based on empirically-based research' (Groat, 1995b: 3). He approached place through cognitive theory, 'which focus[es] upon the links between mental processes (such as perception, memory, attitudes, or decision-making), and social behaviour' (Scott and Marshall, 2005: 92), and on spatial experience, 'hierarchy and differentiation' as well as their implications for the 'content and structure of the conceptual' space.

Canter, a psychologist, sees place as a 'technical term' and considers Relph's notion of place to be 'romantic'. Relph, a phenomenologically oriented humanistic geographer, values authenticity and the particularity of specific places (Gustafson, 2001: 6).

Places are not abstractions or concepts, but are directly experienced phenomena of the lived-world and hence are full with meanings, with real objects, and with ongoing activities.

(Relph, 1976: 141)

Phenomenology emphasises the body sensory experience and emotional relationships

> *Relph's model, on the other hand, was developed through 'phenomenological perspective in human geography' (Groat, 1995b: 3) that considered the experience/phenomenon of the lived world rather than 'theories and abstract models of place' (Relph, 1976). Accordingly, I deduced the model and its diagram from the literature. He considered place identity – developed from Kevin Lynch (1960) – as the 'basic feature of place experience' of both place and people. Place identity for Relph involves 'the recognition of differences and of samenesses ... [and] sameness in difference' (Relph, 1976: 45).*

For David Canter place is a unit of environment and a consensus of cognitive maps, apparently without history; he believes that through our conceptual systems places can be measured and this information then used to make better places. (Relph, 1978: 237)

> *For me places are particular settings with their own history and aesthetic properties, that have personal and communal significances, and which elude measurement.*
> *(Relph, 1978: 237)*

Sime (1985) highlights the opposition between the two models through the people-place relationship, which is subject to scientific measurement in Canter's model and hence differs intrinsically from the phenomenological/emotional sense of place experience.

Canter (1977b: 119) views Relph's model through his own perspective to place, 'Relph chooses to call it a phenomenological perspective, but it could be just as justifiably be referred to as a cognitive approach to human location activity'. He considers the main contradiction between the two models in the definition of the designer/user role towards place, i.e. the author/reader role as elaborated in 'A turning point'. Canter emphasises the role and perspective of the user in his model as will be further explored, whereas Relph considers the designer's perspective (Canter, 1977b). Relph's emphasis on the designer's view was criticised because it was difficult 'to know whether the views of an author are a true reflection of the people whom he or she is referring to' (Sime, 1985: 32). Relph, like Sime, emphasises the differences between place as a measurable unit in Canter's model and lived experience in his model.

Model statement and diagram

> [P]lace is the result of relationships between actions, conceptions and physical attributes. It follows that we have not fully identified . . . place until we know:
>
> - what behaviour is associated with, or it is anticipated will be housed in a given locus
> - what the physical parameters of the settings are
> - the description, or conception, which people hold of that behaviour in that physical environment.
>
> (Canter, 1977a: 158–159)

Hence place is presented at the intersection relation (Figure 5.2). Canter's (1977a: 159) representation challenges the notion of a predetermined starting point, 'we can proceed with the identification of places starting with any of the major constituents'. Canter also considers the possibility of approaching place through either physical attributes or activities and the related conceptions which would help define

▶ **FIGURE 5.2**
Canter's model of place (Canter, 1977a: 158)

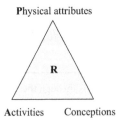

Physical attributes

R

Activities　　Conceptions

R= Relations at intersection

the other elements, activities, or physical attributes.
In other words, place definition is approached
through two elements, taking into consideration
that conceptions are involved. However, Canter's
emphasis on the user perspective (people), and thus
the meaning and conceptions they give to place,
'actions, or behaviour are an essential component
of place and therefore meaning' (Groat, 1995b: 4),
which implies a hierarchical approach to place.
People, as represented through their activities in
place, are considered first and the physical setting
of place is approached last. This hierarchy and its
implications are discussed further in the study of
the model elements and interrelations.

Relph's model 1976:

The identity of place is compromised of three
interrelated components, each irreducible to the
other:

- physical features or appearance
- observable activities and functions
- meanings or symbols.
(Relph, 1976: 61)

Relph considered the three elements as 'distinctive
poles' which are 'interconnected' within place
experience and constitute other internal-divisions.
As Relph himself was not concerned with theories
and abstract models of place, I thus constructed
an abstract model to represent and illustrate his
approach to place (Figure 5.3). The three elements
are represented through three interlocking elements.
He considers the interrelations (Rs) between the

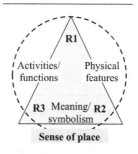

◀ **FIGURE 5.3**
Relph's model of
place

place components and presented a short account of these interrelations which will be explored further in the model elements and interrelations. However, Relph's model involves a fourth, less tangible element which he calls 'sense of place'. The latter helps to interconnect the other elements. However, it exists independently from the other three elements and any changes that occur in them. Finally, it should be taken into consideration that Relph's model only represents a part of his approach to place and place identity, which involves three other aspects as well. The model of place represents the components of 'the identity of place'. The second considers 'the identity with place', 'forms and levels of insideness and outsideness'. The third considers the relation between people (individuals/groups) and place identity. The last considers the development and change of these identities (Relph, 1976: 46).

In between the models' similarities and oppositions, Canter's model is considered to be a reflection on place experience in the intersection between the three elements inside the model. Relph's model on the other hand, approaches place through a multi-levelled framework. The fact that the model represents only a single layer questions the criticism that Relph's phenomenological approach isolates place from the wider socio-culture context (Canter, 1977b; Sime, 1985); this is debated on a different level, outside the model as presented here.

Author/reader

Designers are, officially, the modifiers and creators of physical forms . . . for specific activities and conceptions.
(Canter, 1977a: 161–163)

Through his approach to place, Canter developed a user-based model to evaluate place. Canter considers the evaluation of place as a relationship within the 'conceptual system' between the client, user, and designer, i.e. the third component is represented as client intentions, or organisational objectives.

This model representation is similar to Vining and Stevens's (1986) model represented in 'A turning point' interlude. Both models involve the designer and the user (public). However, place (physical attributes), which represents the third component in Vining and Stevens's model, is not included in Canter's model as will be discussed.

On the other hand, the relations in Canter's model are represented by a circle rotating in one direction. This implies a sequential relation between the components, where every component client, designer, user influences the following one. Canter considered the client's intentions as the starting point, the organisational objectives. He also questions the identity of users, identified as the public by Vining and Stevens, as members of the public or other professional bodies using place.

> In Vining and Stevens's model the relations between the components consists of two circles that rotate in opposite directions. This helps to construct mutual relations between the model components that have no identifiable start point.

The designer's role is to understand the 'ambiguity of place' through the user's (individual(s), group(s), professional bodies) reading of place. Accordingly, he defines the component of place 'to convert this conceptual system into a product' through understanding the user's reading (Canter, 1977a: 164). An illustration of Canter's approach to place discourse is represented in (Figure 5.4). Canter's emphasis on the user's perspective has helped to distance place, the physical attributes, from the discourse. Furthermore, the designer role is confined to a second reader of place and people-place relation through people, the users.

> [T]he role of the architect and his responsibility to interpret, and understand the culture and change the society through his architecture. (Abdelhalim, 1989: 236; Abdelhalim, 1996)

> Finally, designer/user is represented by Vining and Stevens as author and reader (observer) respectively, whereas Canter emphasises the role of the user as part of the conception system of place, i.e. both author and reader.

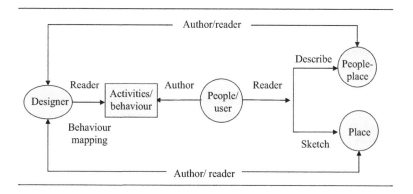

◀ **FIGURE 5.4**
An illustration of Canter's author/reader role

The task of the designer is to disentangle these containers of order and discover their underlying geometry.
(Abdelhalim, 1988: 143)

Relph, on the other hand emphasises the role of the designer towards place. However, he refuses the confinement of the designer's role to management and manipulation of the physical attributes of place, as well as 'the possibilities of place-making on behalf of people' (Sime, 1985: 35). Accordingly, he denies the people's role as author/reader in the discourse of place.

In contrast to Canter's approach, this book is concerned with the designer's role towards the urban space, although I have the same understanding of the designer's role in reading the people, and the people-place relationship is maintained. However, the emphasis on the designer's role differs from Relph's, who considers the designer as an expert and excludes the user from his model. My reading is interested in the designer's reading, interpretation, and making of urban space without excluding the user's role, i.e. highlighting the designer's role towards urban space rather than eliminating the user's role.

Elements and interrelations

As previously discussed both Relph's and Canter's models are made up of three elements which are abstracted through the primary framework of interpretation: place (physical attributes and feature), people (activities and function), and people-place relation (conceptions, meaning, and symbolism). In addition, Relph's model involved a fourth element, the 'sense of place' which represents a relational link between the elements. However, 'the concept of place' differed with Canter's 'objective measurement' of the relation between people and place, and Relph's 'humanistic' approach to experience. This section explores the definition of the elements as well as the internal structure that builds up the relations in each model.

Canter's model is criticised for the lack of adequate definition of the elements, especially the physical attributes and actions. The definition given by Canter of physical attribute is rather a definition of the relation between physical attributes and 'other components of the place in question, those which facilitate the identification of places' (Canter, 1977a: 159).

Relph, on the other hand, defines the model elements as 'distinctive poles', considers the interrelation between these elements and introduces a definition of these relations.
(Relph, 1976: 47)

Place

Although Canter identifies the element of place as the physical attributes associated with the concept of place, he is criticised for not providing details of the physical form (Groat, 1995b; Relph, 1978; Sime, 1985). He considers the construction of this physical space as the task of the architect/designer. In reality Canter considers the definition of physical attributes in relation to 'psychological and behavioural process' (Canter, 1977a), i.e. in relation to people.

Groat and Després (1991) introduced 'five major properties' of place, which are central to the architecture discourse and could be integrated through the social studies of place. These are:

- the architectural style
- composition (geometry, hierarchy, and proportions)
- type (formal and functional structure)
- urban morphology (spatial relations)
- place.

Relph, on the other hand, acknowledges the architect and urban designer professional perspective of place. Accordingly, he is criticised for this subjective approach which does not consider the people's experience in place (Canter, 1977b; Groat, 1995b; Sime, 1985).

Place represents an inclusive phenomenon that could not be reduced to any of the other properties. Furthermore, Groat and Després (1991) reflect on the tendency of social studies to involve architectural style and type and exclude composition, place, and especially urban morphology.

Canter's and Relph's approaches to define the element 'place' in/without relation to people are thus considered a main difference between them. Relph also does not draw enough attention to the physical form. His approach to place considers the 'appearances' of the built environment and nature, i.e. the architectural image and styles. Sime (1985) points out that both these models, as with other social studies of place, have generally lacked an adequate description of the physical attributes of place. Groat (1995b) reflects on the importance of including a well-developed account of the physical attributes into the model of place.

People

As previously discussed both models consider people's activities and behaviour as associated with place:

Canter emphasises the user's perspective. However,
as in the case of physical attributes, he does not
provide a detailed account of the possible patterns
of people's actions (Groat, 1995b; Sime, 1985).

In between the 'community' and
'child' lies a conflict . . . about the
role of the park
(Chapter 4, this volume)

*Simultaneously, Canter criticises Relph's emphasis
on the architect/designer perspective, 'activities
and meanings cannot be driven directly from
the lines on a master plan or details of building
specifications' (Canter, 1977b: 120). Relph
considers people's activities through two levels
of interpretation. The first level considers the
characteristics of these activities in relation to
place, creative, destructive, passive. The second
considers the people who produce these activities as
individuals, and groups (Relph, 1976).*

People-place: meaning/conception

In between people and place lies the meaning and conception given by people to place through the models of both Relph and Canter.

> In architecture . . . the building or its formal details (as the signifier) stand for
> certain meanings or concepts which are the signified.
> (Groat, 1981: 75)

In the traditional paradigm of place
the mind is defined as logic and
rational . . .

The people–place relation in Canter's model is
scientifically measured; it represents a rational
relation following the traditional paradigm.

. . . whereas it is considered as both
rational and emotional in the new
paradigm.

The meaning of places may be rooted in the
physical setting and objects and activities,
but they are not a property of them – rather
they are the property of human intentions and
experiences.
(Relph, 1976: 47)

*Relph's phenomenological approach represents an
emotional relation, a stage between the traditional
and the new paradigms of place.*

Groat (1981) has demonstrated how both Canter's approach and social studies of place in general have emphasised meaning, the signified, over the physical form, the signifier. On the other hand, 'architectural theorists' have tended to focus on the physical place rather than the meaning (Groat and Després, 1991).

Sense of place: a fourth element

There is another important aspect or dimension of identity that is less tangible than these components . . . yet serves to link and embrace them: This is the attribute of identity that has been variously termed 'spirit of place', 'sense of place', or 'genius of place (genius loci)', all terms which refer to character or personality.
(Relph, 1976: 48)

'The sense of place' is the fourth element included in Relph's model, but not considered essential in Canter's model (Sime, 1985). A basic definition of this term is 'the ability to recognise the different places and different identities of a place' (Relph, 1976: 63). Sense of place involves but is not confined to:

topography and appearance, economic functions and social activities, and particular significance deriving from past events and present situations.
(Relph, 1976: 48)

The architect . . . emphasises the value of the history of Al-Sayyida Zeinab, the community draws strength and pride from its reservoir of history.
(Abdelhalim, 1996: 54; Chapter 4, this volume)

Accordingly, the sense of place acts as a relational link that helps the interrelation between the elements of the abstract model. It also helps to relate this abstract model to the wider socio-cultural and economic context. However, the phenomenological perspective is criticised for its separation from this socio-cultural and economic context (Groat, 1995b). It might be interesting to consider the development and integration of the concept of the sense of place as a relational element within the primary framework of urban space to help locate it within its wider context.

Model of interrelations: intersection vs. union

Canter's place is represented at the intersection of the Venn diagram (Groat, 1995b) (see Figure 5.2). Relph (1978) criticises this intersecting interrelation for blurring the definition of these elements, and accordingly their relations: 'environment, building

In the new paradigm:

the interrelations . . . 'embodied and dispersed'
between [non-autonomous] spheres.
(Chapter 2, this volume)

and even people could be substituted for places'
(Sime, 1985: 34).

The traditional 'separatist' paradigm held [place constituents] as autonomous independent spheres. . . . The interrelations between these spheres follow a direct and linear method; an organised internal structure. (Chapter 2, this volume)

Relph, on the other hand, considers the elements of place as well-defined 'poles' that are interlocked with each other and 'interwoven' through the place experience. Relph thus presents a short account of these relations which are well-defined as R1, R2, and R3 (see Figure 5.3). The first relation (R1) between the physical attributes of place and activities draws functional territories and relations similar to the 'functional circles of animals'. The second relation (R2) considers 'the landscape/townscape experience' that develops between the meaning and the physical setting. Finally, the relation R3 between the people's activities and meaning involves the history, culture, and social aspect of place.

THE FACET THEORY OF PLACE

Canter's early model of place discussed in the previous section, was continuously criticised for overlooking the physical environment, and was urged to integrate with architecture theory (Groat and Després, 1991; Sime, 1985). Accordingly, Canter (1997) attempted to respond to this criticism through the integration with architecture theory, by involving Markus's trilogy to 'facet theory', a structural meta-theory and hence introduced the 'facet theory of place' (Figure 5.5). The significance of this theory, besides its richness as it draws on three different disciplines, lies in its integration into architecture space and the use of the facet theory. However, its main disadvantage as expressed by Canter himself, is the need for professional knowledge or expertise in order to understand and work with it (Canter, 1985). In addition, this theory was tested through a number of empirical studies which, while helping to emphasise some positive aspects of the theory highlighted other conflictual aspects that require further exploration.

▶ **FIGURE 5.5**
The setting of the facet theory of place

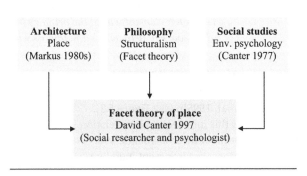

The facet theory is a meta-theory like deconstruction. Accordingly, the projection of this theory through the primary framework of urban space is expected to be difficult. A facet is not an element, but a set of elements. The reading of the facet theory of place thus involves a reverse approach to that of the Canter and Relph models in the last section as it explores the facets of place and their interrelation and then maps the theory (Abdelwahab, 2012). Consequently, deconstruction is included in conversation with the facet theory in order to further explore and reflect on facet theory rather than deconstruction. The latter was thoroughly discussed in Chapter 2. At the same time, the reading of Markus's architecture space is included in the following chapter.

What is facet theory?

> The facet approach was developed by Guttman (1965), described in detail by Borg (1978) and Shye (1978) [behavioural sciences], and reviewed in relation to applied psychology by Canter (1982).
> (Canter, 1983a: 672)

The facet theory helps to address the multivariate nature of social space through a systematic approach (Brown, 1985; Canter, 1997). Canter's argument is that the empirical approach in psychology, and social studies in general, helps to relate the research to quantitative methods; qualitative data were often overlooked because of the analytical difficulty and lack of systematic results (Canter, 1985). Accordingly, a facet approach could help in the effective analysis of qualitative data without compromising empirical data. He explains the application of this approach on two levels:

- **data construction** – the definition of the research facets: the categorisation of qualitative data in a way that helps to embrace the complexities and dynamics of social research
- **framework of interpretation** – the provision of appropriate analytic process: 'a content analysis framework' that does not require 'high levels of measurement' (Canter, 1997: xi).

> *What facet theory is not: deconstruction.*

> There is no set of rules, no criteria, no procedure, no programme, no sequence of steps, no theory to be followed in deconstruction.
> (McQuillan, 2001: 4)

> Choosing the facet approach requires a shift in
> thinking, an imaginative leap even, not only in the
> conception of the research problem but also in
> the design and execution of the inquiry.
> (Brown, 1985: 17)

The facet theory hence, according to Canter (1983a:
672) is made up of three parts:

- 'A formal, detailed definition' of the research
  subject
- 'Empirical evidence, . . . observations'
  concerning the research subject
- And a 'rationale', the logic that relates these
  observations to the definition.

Author/reader

> The value of the facet approach derives from the
> fact that it provides metatheoretical framework
> for empirical research
> (Canter, 1983b: 35)

Canter emphasises that the facet theory is not
an explanatory theory but a meta-theory, which
only provides a structural framework to help
formulate and evaluate theories. Accordingly,
the facet approach does not favour a theory or
an academic position but works from within the
defined facets, or data categories, 'to generate
their own theoretical frameworks' (1983a: vi).
At the same time, Canter (1997) developed the
'facet theory of place' which addresses the role of
architects and urban designers, and is interested
in place evaluation, particularly 'Post Occupancy
Evaluation' (Donald, 1985: 173). However, as was
discussed earlier, the facet theory of place remains
ambiguous to many readers; it requires professional
knowledge of facet theory.

> Deconstruction is not a method . . . that
> is applied from outside the discourse,
> deconstruction takes place within.
> (Lucy, 2004; Chapter 2, this volume)

> In other words, deconstruction does not
> present a 'systematic and closed' procedure
> to reach the meaning.
> (Royle, 2000; Chapter 2, this volume)

Meta-theory is 'a theory the subject matter of which is another theory'.
(Encyclopedia Britannica)

Both facet theory and deconstruction are meta-theories that approach the development of other theories, e.g. place, Khōra. However, facet theory involves a structural approach to categorising the order and hierarchy of an expected well-defined subject. Deconstruction on the other hand, 'wishes to undo . . . structures of all kinds' (McQuillan, 2001: 12). It works as a viral form inside the not-well-defined to destabilise the structure through différance, 'the systematic play of differences' (Derrida, 2004 [1979]: 27), which entitles the structure to continuously defer in space and time as soon as it is established.

What is ~~an element~~ a facet?

A facet is any conceptually distinct way of
classifying the universe of observations.
(Canter, 1983a: 673)

One main difference in the facet theory of place is
the use of the facet instead of the element to describe
place components. A facet is not an element. On
the contrary, it constitutes a set of elements, which
are related and grouped according to the same rule
(Levy, 2005). 'A facet is any conceptually distinct
way of classifying' the different variables of the
research object (Canter, 1983a: 673). The facet is also
different from an element because it is a method
of categorisation (Canter, 1983b). A facet helps to
group similar elements and their definitions and
accordingly, to understand their internal structure
and relations. According to Canter (1997), each facet
should include all sub-elements of the object studied:

* these sub-elements should be exclusive to their
  facet set
* each facet set involves multi-faceted
  categorisation; facet groups and elements
  should be well-defined by the researcher.

What is a ~~facet~~ 'supplement'?

The supplement would be a sub-element at the
centre of a set, which instantly defers to something
outside the set. Accordingly, it is substituted by
a trace of other elements as it breaks outside the
structure, the exclusivity, of the facet.

~~Elements~~ Facets

Canter's facets of place are a development of his
earlier model in 1977. As discussed earlier, the 1977
model of place constitutes the intersection between
the physical environment, people's activities, and
the conception they hold for place. The facet theory
of place involved Facet D: aspects of design; Facet
A: functional differentiation; and Facet B: place
objectives, which respectively, reflect on and develop
the elements of the earlier model (Gustafson,
2001). The Facet C: scale of interaction is a new
addition that recognises the relation to the context.
Accordingly, this reading will now explore the role
of each facet, constituents, as well as structure (see
Figure 5.6). Simultaneously, a reflexive reading of
these facets is considered through the perspective of
deconstruction, which attempts to read the inherent
instability in the facet theory of place.

▶ **FIGURE 5.6**
The development
of Canter's early
model (1977) to
facet theory of
place (1997)

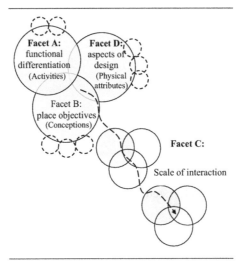

Facet D: aspects of design: architecture space – 'place'

[T]he aspects of design focus on physical
characteristics of place.
(Gustafson, 2001: 6)

In the facet theory, Canter responds to the
criticism of his earlier model of place, which gave
insufficient attention to the components of the

physical environment, and to the difficulty of
integrating his approach with architecture theory.
Canter's response involved the integration of
Markus discourse on architecture space with its
concepts of space, form, and function.

> Canter's definition of place hence adopts the container space, well-defined
> boundaries between the outside and the inside urban space and context
> respectively.

Canter's approach to involving architecture space
continues to acknowledge the professional authority
of architects and designers on place. However, he
confines architecture space to physical space as
well as a subset of his approach to social space. And
the architect's role is to tackle and construct this
physical space (Canter, 1977a).

Canter approaches the architecture space as space-matter 'physical phenomena', i.e. Aristotle's topos.

Facet A: a functional differentiation – ~~'people'~~

> Functional differentiation points at activities.
> (Gustafson, 2001: 6)

Facet A: functional differentiation, does not point
at activities, function, or people as emphasised by
Gustafson. Functional differentiation points at the
differentiation between spaces of function, through
typology of place.

> [T]he hypothesis . . . that has been accepted by
> many architectural theorists . . . that particular
> patterns of activity are associated with particular
> places.
> (Canter, 1997: 131–132)

According to Canter (1997), Facet A consists of
two subsets, central and peripheral. The 'structural
hypothesis' of this facet is that 'a particular
type of place' holds similar/typical functional
spaces, which reflects the central features of a
place type as well as non-similar/atypical spaces,
the peripheral. For example, operation rooms
are central to hospital typology, and waiting
areas are peripheral. The later subset is either
random or structured. Again, Canter prefers the
second type, the structured subset 'that implies

Utilitas (Vitruvius):

Space of Function

↓

Facet A (Canter):

Function in space

↓

(Tschumi):

- space of function body (space-of-movement)
- function in space (body-in-space)
- and between space of body and space of society (event)

Chapter 1, this volume

a structure for the whole pattern of activities' in place (1997: 125).

The functional differentiation follows the modern cause-and-effect: form follows function.

The attempt to define people's activities as independent entities within place has developed three readings of the people-place relation. Vitruvius's definition of 'Utilitas' as the appropriation of space in order to accommodate its function, approached the space of function rather than the function of space. Canter (1997), on the other hand, considered the definition, in terms of differentiation of function in relation to space, central/peripheral. Finally, Tschumi's reading went beyond the function to embody the body-in-space, and the body's spaces-of-movement. He also added another layer through the articulation between space of body senses and space of social context, the event.

Embracing the margin

Deconstruction intrinsically embraces the margin, the peripheral space, which is hence responsible for the structure of place. An architecture metaphor that embraces the peripheral space of function helps to bring about architecture deconstruction beyond stylistic considerations.

Finally, the hypothesis structure of this facet defines and is defined through the other facets of place, i.e. what is central and what is peripheral in relation to place objectives or aspects of design. The structural relationships between place constituents, physical and social space: people-place relation. Simultaneously, this relation constitutes a 'gradation between central and peripheral', which by definition blurs the distinction between the two sets. However, 'the very logic of differentiation facet leads to hypothesis of it being simply ordered' (Canter, 1997: 127).

The cornerstone

Deconstruction does not reverse the binary: centre/ peripheral, which would only develop another binary. The peripheral is displaced through the cornerstone that is responsible for the construction/ deconstruction of the entire structure on the margin of place rather than the centre.

Facet B: place objectives – ~~'people-place'~~ 'people'

> The facet of place objectives has some similarities with the 'conceptions' [people-place] component of this model, but clarifies and extends it substantially by explicitly considering individual, social, and cultural aspects of place experiences.
> (Gustafson, 2001: 6)

Facet A, functional differentiation of place, does not reflect on people's activities, nor their abstraction as 'people' in the primary framework of urban space. Controversially, Facet A reflects on the nature of people-place relation as central/peripheral through a cause and effect relation. Consequently, Facet B, place objectives, is expected to explore the place-people relation through experience. However, Facet B reflects on the categorisation of people's experience in place, and particularly the categorisation of people through the individual, social, and cultural. Canter (1997: 126) identifies this facet through Saegert and Winkel's (1990) review of perspectives of place effectiveness, through the 'different types of objectives' of individuals, social and cultural. Canter also identifies the range of relations between these elements as psycho-social, socio-cultural, and psycho-cultural respectively (Figure 5.7).

The Cultural Park for Children

The manager's emphasis on the child reflects the individual objectives.
The architect's emphasis on the community and the cultural significance of the history of the neighbourhood on the one hand and marginalisation of the child as symbolic on the other, is a reflection on socio-cultural objectives of the park and subversion of the individual objectives including psycho-social and psycho-cultural objectives as symbolic.

> Individual: child
> Social: community
> Culture: history and heritage

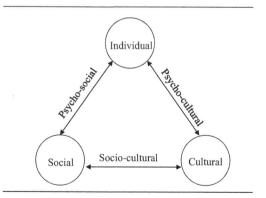

◀ **FIGURE 5.7**
Facet B, set of 'people': individual, social and cultural and relationships

In earlier studies, Canter considered place objectives that lay between the social and physical rather through this categorisation of people (refer to Canter and Kenny, 1981; Canter, 1983a; Canter and Rees, 1982). In other words, he considered people-place relations between Facet D, design aspects, and people, individual, social, and cultural.

Both Sixsmith (1986) and Gustafson (2001)
introduced similar three categories of people
relationship to place through empirical research.
They considered these categories as personal – the
self, social – the others, and the physical – the
environment (Figure 5.8).

At the same time, Markus defines 'three levels of relationship' to identify the
social structure: the self, the others, and the cosmic (Markus, 1982a).

▶ **FIGURE 5.8**
Levels of meaning
according to
Gustafson (2001),
Sixsmith (1986) and
Markus (1982a);
diagram developed
for Gustafson

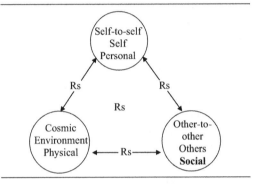

For Markus (1982a: 6) the first level constitutes a relation of 'self-to-self'
that holds three basic questions, 'who am I?, where am I going?, what am
I becoming?'.

Sixsmith (1986) considers the personal as both the
self which is grounded in the person's emotions,
desires, and activities, and the being which
considers the person's identity through place
attachment and belonging. Gustafson (2001)
also considers the self as grounded in personal
meanings: emotions, activities, and self-identity, as
well as life experience and memories.

Markus's second level constitutes the relations of 'self-to-others' as well as 'other-
to-others', where this other oscillates between divine and reason, science or society.

Cairo space
Individual: Cairene/tourist
Social: . . . .
Culture: history and heritage

The social for Sixsmith (1986) resembles Markus's
idea, i.e. the social constitutes the relations of

'self-to-other' and 'other-to-other' and represents the social function of place. Gustafson (2001) further explains 'self-to-other' relations as a 'sense of community' which considers the recognition and anonymity of the self to the other(s). However, he defines the other through the difference between 'us-here' and 'them-there'.

Markus's third level considers the relations of 'self-to-cosmic' and 'others-to-cosmic' in which the cosmic order is built up of myth, philosophy, religion, as well as the city's physical presence and geometry.

Sixsmith (1986), on the other hand explores the physical place as involving space 'structure and architecture style' in addition to people's space; it also involves place 'services and facilities'. For Gustafson (2001) environment involves both the physical and natural environment, as well as the institutional (political). He takes into consideration the 'distinctive features and events' that help to develop the identity of place (e.g. place types) on the physical level, as well as on the symbolic (e.g. historic). Accordingly, he examines the relationships between the self-environment and the other-environment. He also considers the self-other-environment relationships through citizenship and similar memberships to institutions and distinctive places. Finally, it should be taken into consideration that the social perspective towards place meaning is variable dependent on different individuals, and groups as well as the place; i.e. not all meaning categories and subcategories are relevant in every situation (Gustafson, 2001).

Both Markus's definition of meaning and other empirical research considered similar categories of meaning, the self, the other, and the place. However, Markus, working within an architectural background, viewed meaning from a more abstract/theoretical perspective, questioning the self/being, the divine/logic other, the myth and philosophy of place; the latter however, was seen as embedded within the physical presence of the city but without details.

A simple order of relationships from psychological through social to cultural would not be found.
(Canter, 1997: 127)

Saegert and Winkel (1990) proposed a distinctiveness between individual, social, and cultural objectives of place. This distinction reflects a 'different aspect of a coherent system of place' (Canter, 1997: 126). However, Canter also considers their simultaneous occurrence. The distinction between these categories hence refers to the 'emphases' of the different place to these objectives and thus 'gradation may be identified between the three elements'. Canter reflects on the sequential order of this set, individual, psycho-social, social, socio-cultural, cultural, and psycho-cultural (see Figure 5.9). This latter relation, psycho-cultural, destabilises the proposed hierarchal order as 'the sequence would have to double back on itself to provide a position for an element between the supposed two extremes' (Canter, 1997: 126).

A sequence is 'a number of things, actions, or events arranged or happening in a specific order or having a specific connection' (Encarta-Online-Dictionary). This definition implies a linear hierarchal order that has a start and leads to a particular point.

▶ **FIGURE 5.9**
Gradational relation between the elements of people's set, Facet B

Facet C: scale of interaction – 'context'

[T]he facet of scale of interaction also adds to Canter's earlier framework by pointing out the importance of environmental scale.
(Gustafson, 2001: 6)

[C]oncept, content, context . . . for Tschumi context . . . includes the 'historical, geographical, cultural, political or economic' urban context.
(Chapter 1, this volume)

Canter brings the categorisation of the setting, the context, to the discourse of place which 'is often ignored in theory building'. Facet C, the scale of interaction, is a quantitative set that considers the 'difference between uses of space' at different

context levels: immediate, local, and distant
rather than studying the context itself (Canter,
1997: 127–128). Canter also discusses the relation
between Facet C and Facets A and B; i.e. context
and place. He considers two alternative hypotheses:
the interdependency between place and context
on the one hand, and the independency of context
on the other.

The manifestation of place-context relation as
'interdependent' makes it possible to understand
the context 'from analysis of . . . sub-places'
(Canter, 1997: 128). The independence of the
context however complicates this analysis. Each
level should be addressed as a holistic experience.
However, Canter states that the independence of
the context also denies the differences between the
different levels; 'no matter what the scale of the
place, the personal, social, and cultural elements
would be identifiable in much the same way'
(Canter, 1997: 128).

Cairo space
Immediate: local
Local: regional
Distant: global
The Cultural Park for Children
Immediate: the park
Local: Al-Sayyida Zeinab
neighbourhood
Distant: Cairo

'Il n'y a pas de hors-texte' . . . 'It is worth noting that the blurriness in the
text between content/context and text/author does not imply a homogenous
production of truth but rather helps to embed and recognise the multiplicities
within the text that are inscribed in the context'.
(Chapter 2, this volume)

For deconstruction, the context-place relation(s)
are both interdependent and independent. The
distinction between the content (place/text) and the
outside context (metaphysical, historical, social,
and so on) is denied. The content holds a trace of
the context, which is hence inscribed inside the
place. At the same time, the context traces place.
The blurriness of the content of place, inside/outside
holds another implication for place which also traces
and is traced by the author – designer – together
with the context.

The facet theory of place brings in another perspective of context: the nature
or type. This approach goes beyond the binary representation of place, e.g.
a local/global reading of Cairo space, as it explores various levels of reading:
local, regional, and global. This perspective both questions and helps the co-
existence of these levels, these readings of place.

Social space projection mapping sentence

> The mapping sentence . . . coordinates formal
> concepts (facets) and informal verbal connectives.
> (Levy, 2005: 179)

As part of the facet approach, Canter presented his
theory of place both verbally through a 'mapping
sentence', and graphically through a '3D cylindrex
model of place' (Figure 5.10). The mapping sentence
is a basic device of facet theory. It provides a 'formal
definition' to explain the main elements of the
research objects, grouped together into facets as well
as explaining internal relations within the facets and
between them (Canter, 1997, 1983a; Brown, 1985).

▶ **FIGURE 5.10**
Cylindrex model of
place (evaluation)
(Canter, 1997: 139)

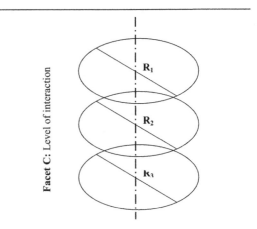

Facet C: Level of interaction

R_1

R_2

R_3

Rs: Relations between Facet A and Facet B

The mapping sentence of the facet theory of place
includes:

- a study population which considers the
  architectural design aspects: Facet D
- a domain-content that includes functional
  differentiation, objectives and scale of
  interaction of place: Facets A, B, and C
- and the common range of evaluation of place
  effectiveness (Canter, 1997).

The relations between these elements and facets
are shown though connecting words but without
the provision of details (Canter, 1997). A relation

is built through 'one and only one' from the range facet for each generated domain sentence (Levy, 2005). The aim is to provide a number of hypotheses to be tested empirically (Canter, 1997). Though the mapping sentence explained previously implies a network of complex relations, the working relation between the elements carries a linear potential through choosing one and only one range element to describe the chosen element of the other facets. These relations are built up through the relation of one element to another in another facet and so on, in a linear fashion; linear equation can be safely used (Canter and Kenny, 1981). This linear reading of place will be re-approached in Chapter 6 in a reflection on a reading of architecture space.

The content-domain, facets A, B, and C, is represented through a cylindrex model (Canter, 1997). The cylindrex structure has shown a 'remarkably consistent occurrence' in the repetitive experiments through other theories, methods, and observations (Canter et al., 1980 cited in Canter, 1983b: 58). A cylindrex model of place is a hypothesised three-dimensional representation of the facets, the sets of elements, and the structure of the facets. It is composed of a place recurring along the axis of the cylindrex. It is built up through the relation between Facets A and B, people and people-place. The model is derived from the fact that 'the facet . . . does have the interesting property of being ordered', i.e. structured (Canter, 1983b: 56). Different types of facet order are formed through different facet studies, while order is produced through one facet modifying another. This order implies dependency between the facets. The axis of the cylindrex, on the other hand, represents the levels of interaction, categories of context, Facet C, which being independent from, and non-planar to Facets A and B, forms the axis of the cylinder.

A reading of 'The Cultural Park for Children' using the facet theory of place is demonstrated in Figure 5.11. A reading hypothesis would involve the study of effectiveness of the 'form' of the park in terms of the child's central activities and psychology and in relation to the Sayyida Zeinab community, which itself is a reflection of the manager's reading. Another hypothesis would involve the study of the effectiveness of the park morphology and organisation as central to child activities and as satisfying the social and cultural objectives of the park.

▶ **FIGURE 5.11**
Facet theory
mapping The
Cultural Park for
Children

Facet D
The extent to which aspects of design of place (p) achieves

[1. Function: architecture
programme]
[2. Space: morphology and
organisation: Ibn Tulun
Spiral Minaret – and
festival rituals]
[3. Form: geometry and
style – nature (palm trees)
and Islamic features]

Facet A Differentiated	**Facet B** Place objectives at	**Facet C** Scale of Interaction
[1. Central: child/ community] [2. Peripheral: community/child]	[1. Individual psychology: child] [2. Social: community] [3. Cultural: history and culture Monuments and festivals]	[1. Intermediate: park] [2. Local: Al-Sayyida Zeinab] [3. Distant: Cairo]
Will be →	Common range [.Effective] To [.Ineffective]	Achievement of objectives through design aspects

Where place (p) is one of a population (P) of places that are experienced by
people and open to empirical study

ON THE MARGIN

This chapter re-approached social space, through two reflexive instances: the reading of Canter's and Relph's models of place, and the facet theory and deconstruction meta-theoretical approach to place. In the first instance, Canter, on the one hand, adopted a cognitive approach to place that emphasised the rational 'mind' relation to experience of place and social behaviour and psychology, and Relph on the other assumed a phenomenological approach that drew attention to the 'body' and its emotional experience in space. However, both models could not present an adequate definition of physical space. At the same time, their approach to place reflected a positivist, well-defined reading through linear and direct relations that overlooked the dynamics and complexities within.

In the second instance, both facet theory and deconstruction recognised the dynamics and complexities of place and the difficulties to approach the qualitative date in particular, which required a shift in the way of thinking and understanding of place as well as 'how' to approach place. Therefore, both approaches attempted to work from within place to explore it. However, facet theory of place works through a pre-defined framework and system of categorisation to structure the reading of place – the facet – whereas deconstruction attempts to destabilise the inherited structural representation through the supplement that escapes these categories and facets.

At the same time, the facet theory of place attempted to respond to the criticism to Canter's earlier model that excluded architecture space, particularly the physical space, through the inclusion of Markus's discourse on architecture space. However, this inclusion subverted architecture space as a subset of social space, Facet D: the aspects of design. This subversion implicated both the social and the architectural space. It primarily confined architecture space to the physical, and misread Markus's approach which, as demonstrated in this chapter and the following, extends to include the social space. In addition, Facet A: functional differentiation, echoed Vitruvius's Utilitas, Markus's 'function', and Tschumi's sequential triad, space, movement, and event. Facet B: place objectives, which categorises the 'people', copied Markus's reading of 'people' in relation to place through the self, the other, and the context. Finally, Facet D: the scale of interaction, endorsed Tschumi's reading of the context as well as the milieu, the figure of architecture landscape in Khōra.

Eventually, the facet theory identified and represented the complexities of place through the sophisticated mapping sentence and the figure of the cylindrex. However, it approached the reading of place through a linear and direct relation built through 'one and only one' element of a facet in relation to another only one element in another facet, i.e. the facet theory of place represents multiple readings of place, where each could be read through a linear equation in order to study the range of its effectiveness and ineffectiveness. This reflexive reading of social space thus extends us to re-approach architecture space in the following chapter.

Chapter 6

ARCHÉ-URBAN SPACE

This chapter explores the last reflexive reading of architecture space, through Markus and Tschumi (1980s) in reflection to the facet theory and deconstruction meta-theories, as well as the review on Cairo-Khōra and the vignette 'The Cultural Park for Children' on the margin. In between Markus and Tschumi, as previously discussed, architecture space reflects a paradigmatic shift in the reading of space/place from a traditional notion of independent constituents, which held linear and direct relations in between them, to a new paradigm where these constituents are not autonomous but oscillate between each other as well as being embodied and dispersed in multiple and complex relations between them. This chapter will explore further the background of the Markus and Tschumi approaches to architecture space. On the one hand, Markus approached it in close proximity to language and discourse analysis (Markus, 1987). On the other hand, Tschumi worked through a post-structuralist approach which 'echoes Derrida' but also references other authors from philosophy and literature criticism, such as Barthes and Sollers (Martin, 1990: 33). In addition, the reading of the facet theory developed in the previous chapter highlighted both the readings of Markus and Tschumi through physical and functional space respectively.

The reading of architecture space discussed in this chapter builds on the reading of social space in the previous chapter. We identify the first reflexive instant through the main authors of architecture space, Markus and Tschumi. The second reflexive instant examines the commentators on this discourse, and involves two groups, presented in the horizontal box. The first group involves the facet theory and deconstruction – the meta-theories beyond the readings of Markus and Tschumi – as discussed in the previous chapter. The second group hence involves the discourse on social space. Finally, the reflexive reading of architecture space in relation to the reading of Cairo-Khōra is presented on the margin (Figure 6.1).

The reading of architecture space thus constitutes three parts: a reading in between Markus and Tschumi which considers the backgrounds of their approaches, the constituents, relations and interrelations between their readings of architecture space, and lastly reflections on space/place, Khōra and Cairo.

Author 1: Thomas Markus Vitruvius is well and alive	On the margin, Cairo-Khõra Cultural Park for Children
On the margin, Cairo-Khõra Cultural Park for Children	Author 2: Bernard Tschumi Space Movement Event concept content context
Comments from urban theory	

◀ **FIGURE 6.1**
Social space
discourse setting
through a reflexive
reading between
the discourse actors

THOMAS MARKUS

> 'Discourse' as used here includes everything
> said, written or done in the field. . . . 'Doing' in
> Architecture includes all that is designed and
> built.
> (Markus, 1982a: 4)

Markus's account of architecture space is developed
through his approach to discourse analysis with
a special emphasis on building analysis. Markus
defines the boundaries of this discourse through
form, function, and space (see Figure 6.2).

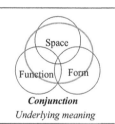

Conjunction
Underlying meaning

◀ **FIGURE 6.2**
Markus's
architecture space

A human language

> The distinction between each discourse is
> not always sharply made – in part because of
> technical difficulties and in part because the
> blurring often helps to maintain the connections.
> (Markus, 1982a: 6)

Markus's approach to discourse analysis considers
both the idea development and end-product, where
the production of the discourse is a social practice,

as previously defined. Markus considers form, space, and function as discourses of 'human language', which consists of vocabulary, length, and organisation of text, points that elaborate on the text and points that are silent. This language signifies the values and intention of author/designer. However, these discourses are sharply identified (Markus, 1987).

Construction of space: the silent points

> Discourse also includes silence – those possible things which are not said, written or done.
> (Markus, 1982a: 4)

In an apparent similarity to deconstruction, Markus reflects on the missing points in the discourse. He defines the boundaries and identifies the missing – silent – points that were not included and then attempts to identify the reason(s) for this exclusion (see Figure 6.3).

▶ **FIGURE 6.3**
Markus silent points: empty slots inside well-defined boundaries (inside/outside)

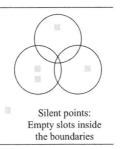

Silent points:
Empty slots inside
the boundaries

However, Markus and deconstruction adopt different approaches to these silent points on the margin of the discourse.

> In order to identify the silent parts of discourse, boundaries have to be defined – otherwise silence is infinite.
> (Markus, 1982a: 4)

Markus considers space, form, and function as a tool for providing a clear analysis of space/place, especially in relation to the values and relations of the social context. However, he also emphasises the missing relation in between this abstract discourse and the context, 'the way in which [space/place] is incorporated into the city' (Markus, 1982a: 5–6).

[O]ne of the terms will be figured within the concerns of the discourse, one as central to discourse, and one marginal; one will be 'included' within the concerns of discourse, one 'excluded' from them. Deconstruction is interested in this so-called marginal term.
(McQuillan, 2001: 23)

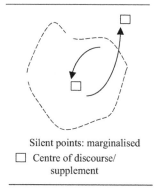

Silent points: marginalised
☐ Centre of discourse/
supplement

◀ **FIGURE 6.4**
Deconstruction silent point:
as the margin/ centre is reversed, supplement escapes the discourse
(~~inside/outside~~)

Deconstruction questions the boundaries and edges, studies the discourse within its context in order to identify the marginalised – the silent points – which exist on the peripheries. The deconstruction hence starts by turning things upside down; the marginalised becomes the centre of the discourse and the centre becomes the supplement which escapes the discourse (see Figure 6.4).

The process of urban space production, 'technology and resources', is not directly included in this discourse though Markus emphasises its relation to the wider socio-economic context (Markus, 1982a: 5–6). He also emphasises importance of 'find[ing] reasons for silent discourse', which he then relates to architectural design and analysis. 'Why in a specific context of space and time, were certain possible things not said, designed or built?' (Markus, 1987: 475).

[Architecture space] . . . may be found in, for instance, philosophy (Lefebvre, 1991 [1974]), new spatial theories (Hillier, 1996; Hillier and Hanson, 1988), politics (Castells, 1977; Harvey, 1973), or in recent explosion of literature in environmental science, ecology, computing and engineering technology.
(Markus and Cameron, 2002: 32)

Although Markus emphasises the well-defined boundaries of architecture discourse, he also acknowledges the blurred boundaries between the discourse on architecture and other disciplines. He highlights the inclusion of architecture space in other disciplinary discourses. Tschumi, on the other hand, involves other disciplines in the construction of architecture discourse, as explored in the following section.

BERNARD TSCHUMI

> I was concerned with the need for an architecture that might change society.
> (Tschumi, 2001: 5)

[W]e were trying to wrestle from the people their authentic culture and adapt it to the expectations of the educated. We wanted to intellectualize it to the expectations of the educated.
(Abdelhalim in Hassan, 1997: 13; Chapter 4, this volume)

Tschumi's early reading of architecture space sought to develop a 'revolutionary' theory (Martin, 1990: 24). However, by the late 1960s, early 1970s, he, together with many architects of his generation, had developed 'a sceptical view of the power of architecture to alter social or political structures' (Tschumi, 2001: 5), a failing that pushed Tschumi to 'put architecture into crisis' (Martin, 1990: 24). Tschumi's readings wandered from Marxism to Lefebvre to a pre-occupation with post-structuralists: Derrida, Foucault, and Kristeva, among others.

[F]ollowing mainly Sollers (limits), Hollier (Bataille), Barthes (pleasure), Kristeva (intertext), Genette (palimpsest), and Derrida (deconstruction), Tschumi introduced into his work the major themes developed by the most visible French literary critics of the 1960s and 1970s.
(Martin, 1990: 33)

The discourse of architecture 'As practice and as theory, . . . must import and export' beyond its disciplinary boundaries (Tschumi, 2001: 17). He thus acknowledges the integrative relations between architecture and other disciplines: 'art, literary criticism, and film theory'. However, his borrowed quotations remain integrated in his text, without quotation marks or author reference. Tschumi hence approached the reading of architecture by drawing on multiple disciplines and through the 'practice of Intertextuality' between disciplines (Martin, 1990: 26). 'Intertextuality' is introduced by Julia Kristeva (1969 [1986]), and considers that all text is embedded in another culture of textuality as it references and quotes other texts (Allen, 2005). Accordingly, Tschumi's account of architecture space developed through the integration of fragments of these readings.

To understand Tschumi's practice of intertextuality, it is necessary to understand his approach to architecture space. Tschumi echoes and simultaneously negates the definition of architecture space highlighted in Chapter 1, which presents architecture space in between mind – knowledge – and the material space, embedded within context. For Tschumi, architecture space is concerned with both the mind and material space. However, both spaces exist simultaneously, and are projected through writing and practising architecture (Figure 6.5). Tschumi's practice of writing on architecture space is simultaneously a practice of writing: knowledge, writing composition, language and style, embedded in between multidisciplines, and a practice of design: concept, physical space, embedded in multidimensional context. Thus, his approach, strategies and tools, and terminologies are projected through his writings – as an approach to writing – his writings on architecture space – as an approach to architecture, his practice – as an approach to making architecture – and his theories – as an approach to reading architecture space.

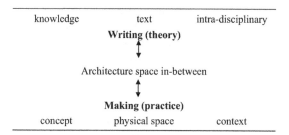

◀ **FIGURE 6.5**
Tschumi's intertextuality: architecture space in between theory and practice

Tschumi's strategies are hence identified through his practice of intertextuality, transgression, and transposition as well as through the anti-synthetic juxtaposition of different layers and fragments; see Figure 6.6. As explained previously, the practice of intertextuality concerns reading between multiple disciplines. Transgression is going beyond the limits of one discipline: theory and practice. However, transgression does not 'destroy' these limits (Bataille, 2001 [1962]; Tschumi, 1976); it provides a 'critique' of 'the boundaries and limits that we construct for ourselves, or that are constructed for us by the dominant power structure' (Hejduk, 2007: 395). Transposition is 'the act of recasting . . . placing in a different setting' (Encarta-Online-Dictionary).

▶ **FIGURE 6.6**
Tschumi's
architecture space
from language
to text

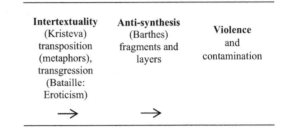

Tschumi simply places concepts, terminologies, and metaphors from other disciplines in architecture,[1] for example: Barthes's *The Pleasure of the Text* (1976), Tschumi's 'The Pleasure of Architecture' (1977) and 'Architecture and Its Double' (1978), and Artaud's *The Theatre and Its Double* (1958; (Martin, 1990; Wong, 2003).

Accordingly, Tschumi borrows fragments of text from outside architecture and integrates them 'violently' into the discourse of architecture (Martin, 1990; Wong, 2003). These layers and fragments are added anti-synthetically 'through intentional juxtaposition and superposition' to the text (Martin, 1990: 30). These practices or reading strategies help the production of space (text/architecture) through the 'deconstruction-reconstruction' of other spaces (disciplines) (Barthes, 1976; Hejduk, 2007; Tschumi, 1976).

This section draws on the Hejduk (2007), Martin (1990), and Wong (2003) approach to reading Tschumi's theory through the analysis of these fragments and their involvement in his work. However, I am not interested in the development of a literature review of Tschumi's work, but in developing an understanding of his account of architecture space. Accordingly, this reading will trace the development of Tschumi's account on architecture space through these fragments rather than studying these fragments in themselves. Tschumi's account is reflected through the architecture paradox – a binary reading – the insertion of a third term – the pleasure of architecture – the deconstruction of architecture space, and questioning the space limits and boundaries (Table 6.1). A full account of Tschumi's displacement of the boundaries of the architecture discourse in the twentieth century is presented in Chapter 2.

Saussure et al. (1955) **Sign= signifier →signified**		Lefebvre (1991 [1974]) **Production of space**		
Barthes (1976) **Literature**	Derrida (1997b) **Philosophy**	Hollier (1992) **literature**	Bataille (2001 [1962]) **Philosophy**	Sollers (1968) **Lit. Critique**
Urban semiology **The pleasure of text** Split between - text and reading - author and reader Insert a third term in between pleasure/eroticism	Deconstruction - **writing/speech** - deconstruction of presence - and binary oppositions - western metaphysics	Metaphor(s) **Pyramid/labyrinth** (No-transcendence)	Architecture **Eroticism** reason/experience	Experience of limits: **Boundaries** - architecture space

Tschumi **Putting architecture into crisis** From language to text Intertextuality: transposition, transgression, Anti-synthesis: layers and fragments *Violence and contamination*

1. Resistance of modernity (pleasure)   3. Practice of intertextuality   4. Research 2. The crisis of the sign   limits
'Pleasure of Architecture' (1977): subverts the transcendence of binary opposition 'The Architectural Paradox' (1975a): from language to text ~~writing~~ architecture	'Architecture and Limits' (1980, 1981a, 1981b) 'Architecture and Transgression' (1976) 'Violence of Architecture' (1981b)

◀ **TABLE 6.1**
Reading Tschumi: intertextual fragments violently integrated into architecture space, based on reading Martin (1990), Hejduk (2007), and Wong (2003)

From language to text[2]

> By focusing on itself, architecture has entered an avoidable paradox that is more present in space than anywhere else: the impossibility of questioning the nature of space and at the same time experiencing a spatial praxis.
> (Tschumi, 1975a: 28)

Tschumi presented architecture space in between (mind), the conceptual 'production of space' – a

Khōra:
a space:
a passage/oscillation
in between
intelligible/sensible
mind/body
being/becoming

transposition from Lefebvre (1991), and (body), the sensory experience of place (Tschumi, 1975a). The contradiction between these two 'interdependent but mutually exclusive' terms emphasised the paradox of architecture space (1975a: 48). Accordingly, he described the two terms as 'parallel' folds detached from the social and economic forces of production; see Figure 6.7.

[T]he concept of space is not space.
(Tschumi, 1975a: 48)

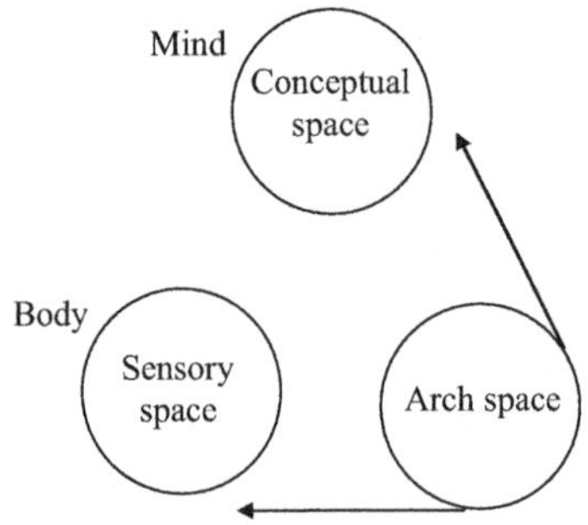

▶ **FIGURE 6.7**
Tschumi's space in
between mind and
body

Plato – Aristotle
Chóra – topos
Space – place
Transcendent – immanent
Intelligible – sensible
Rational – empirical
(Chapter 1, this volume)

To explain the paradox of space further, Tschumi drew on an architecture metaphor used by Hollier (1992) to analyse the work of Bataille (2001 [1962]), i.e. the pyramid and labyrinth (Martin, 1990; Wong, 2003). The pyramid is the symbol of 'reason', the concept of space that 'overlooks' the labyrinth, the symbol of 'perception', the experience of space (Martin, 1990: 27). The paradox of this metaphor is that the concept, the pyramid could not be built to 'overlook' the labyrinth of space. Furthermore, Tschumi (1975a: 49) considers that the sensory experience of space takes precedence over the conceptual. However, there is 'no point of transcendence' in the experience of the labyrinth. 'Bataille was obsessed with architecture' and Hollier's (1992) analysis of his work, the paradox of space, 'established a solid link between text and architecture' (Martin, 1990: 27).

On the margin of the paradox of architecture space, the reading of space developed from a structuralist to a post-structuralist view, 'from language to text' (Martin, 1990: 24). The structuralist concern for language derived from Saussure's binary reading, signifier/ signified and read architecture space as a metaphor of language (Saussure et al., 1955). Barthes (1976) was concerned with architecture 'resistance' to this binary reading (Martin, 1990: 27,24). He developed a link between text and body, and introduced a split between text and reading, author and reader through the insertion of a third term 'pleasure' (Table 6.1). His theory asserted the subversion of the binary reading through 'textual practices' (Martin, 1990: 25). These practices are borrowed in Tschumi's text as elaborated here.

'[T]he rational play of language as opposed to the experience of the senses, would be a tedious game if it were to lead to a naive confrontation between the mind and the body' (Tschumi, 1975a: 43).

De-construction of space: beyond the ~~silent~~

Martin (1990: 31) also highlights another two layers to Tschumi's reading of Sollers (1968 [1983]) and Derrida (1991b). Derrida, as discussed in Chapter 2, was concerned with the deconstruction of the western metaphysics of binary oppositions. He subverted the binary, speech/writing as writing/speech. And Tschumi displaced 'writing with architecture'. Simultaneously, Sollers's work was interested in and entitled *Writing and the Experience of Limits*, and this became a 'central aspect' of Tschumi's theory. Thus, Tschumi continued to experience the limit of architecture space by extending it to other disciplinary spaces. This new experience helped him to displace his early reading of architecture space with space, movement, and event, as discussed in Chapter 1. Consequently, through his reading practice of intertextuality, Tschumi developed multiple accounts of architecture space (Table 6.2). Architecture space is therefore placed in between binary 'alternatives', idea and reality, mind and body, intelligible and sensory spaces (Tschumi, 1975a; Hejduk, 2007; Martin, 1990; Wong, 2003).

'Tschumi's plan is clear. He read architecture through a dualist model and introduced a third term to subvert the duality' (Martin, 1990: 27)

Sensible (body)	In between (third term)	Intelligible (mind)
Labyrinth		Pyramid
	Pleasure	
Program		Space
	Disjunction	
Perception		Conception
	Experience	
Movement		Space
	Event	
Content		Concept
	~~Context~~	

Martin (1990) reflected on Tschumi's reading of architecture space as a binary subverted through the introduction of a third term. This simplified reading could hold the 'cornerstone' that deconstructs Tschumi's architecture space. Tschumi, on the one hand, reads the architecture paradox between two exclusive constituents, and the relation between the triad of architecture space through disjunction. On the other hand, Martin reads the 'third term' as a link that connects the two terms; a relation that explicitly rejects the 'disjunction'. Tschumi's reading of the architecture paradox and the relation between them is hence explored in relation to Martin's reading – potential deconstruction of Tschumi's space.

> [The paradox] . . . always misses something either reality or concept. Architecture is both being and nonbeing. The only alternative is silence.
> (Tschumi, 1975a: 48)

Tschumi (1975a: 44), reads the contradictory constituents of the architecture paradox as 'in fact complementary'. This reading brings architecture space into silence; or the author, in Barthes's text, 'to stop writing, an act that means the destruction of the text' (Martin, 1990: 25). Barthes considers the alternative of introducing 'pleasure', a third term to subvert this binary reading. And Tschumi simultaneously questions the possibilities of 'go[ing] beyond' the paradox: the silence of architecture space (Tschumi, 1975a: 44). Tschumi's approach to this question differs from Martin's simplistic reading in that it involves, but is not restricted to, the insertion of 'pleasure' as a third term. This influences his reading, both the constituents as well as their relations and interrelations.

Constituents – contamination

Architecture was seen as a combination of
spaces, events and movements without any
hierarchy or precedence among these concepts.
(Tschumi, 1989: 255)

Tschumi hence approaches the paradox of
architecture space through 'the imaginary blending
of' the two terms, a blending that brings about the
contamination of the constituents of place as 'It
introduces new articulations between the inside and
the outside.' The boundaries between these exclusive
terms are blurred as the definition of one term
involves the inclusion of the other.

[Khōra oscillates] between two
oscillating types rather than
between oscillating figures. These
two types are double exclusion
(neither/nor) and participation
(both/and). This oscillation denies
polarity and Binary Opposition.
(Chapter 3, this volume)

On the one hand, the experience of space, 'the
labyrinth . . . includes the dream of the pyramid.'
On the other hand, only through the recognition
of the reason, the architectural concept, 'the
subject of space will reach the depth of experience
and its sensuality' (Tschumi, 1975a: 51, 50, 49).
At the same time, the third term, the pleasure
of architecture, is also contaminated. Tschumi
(1977) introduces two types of pleasure, that of
experience of space and that of reason. Neither of
these 'on its own' is the pleasure of architecture
which is intrinsically in between concept and
experience 'where the culture of architecture
is endlessly deconstructed and all rules are
transgressed' (Tschumi, 1977: 85, 92).

Relation(s) and interrelations

Consequently, the paradox of architecture space
and the insertion of a third term helps to break
the pre-established relations in the reading and
representation of space as 'it suggested new
oppositions between dissociated and new relations
between homogenous spaces' (Tschumi, 1975a: 50).

But these texts refuse the simplistic relation
by which form follows function, or use, or
socioeconomic . . . any cause-and-effect
relationship between form, use, function, and
socioeconomic structure has become both
impossible and obsolete.
(Tschumi, 2001: 4)

IN BETWEEN MARKUS AND TSCHUMI

Each of these three experiences of form,
function and space . . . tells us something about
relationships – human relationships.
And they do so in three different ways. They tell
us something about ourselves . . . other people . . .
universal principle.
(Markus, 1988: 345)

If architecture is both concept and
experience, space and use, structure and
superficial image – non-hierarchically – then
architecture should cease to separate
these categories and instead merge them
into unprecedented combinations of
program and spaces.
(Tschumi, 1989: 254)

Markus's and Tschumi's approaches to reading architecture and urban space are a projection of the continuity and disruption, respectively, of the traditions of architecture between neo-Vitruvian and a new paradigm of space/place. However, both readings reveal the dynamics and complexities of urban space. Markus, on the one hand, attempts to develop a peaceful reading, well defined and well structured, of the inherited complexities. Accordingly, he emphasises the distinction between the constituents of his reading of architecture space and simultaneously reflects on the connections between them that blur this distinction. Tschumi, on the other hand, attempts to produce a chaotic reading of the well-defined and structured misrepresentation of architecture space. Accordingly, Markus aims to develop a singular reading that recognises and understands the multiplicities and complexities of space/place through a traditional linear representation, well-defined hierarchal boundaries, categories, and relations. His reading is developed outside architecture space through a transcendent social production. Conversely, Tschumi aims to develop multiple readings of the different instances within the dynamic complexities of architecture space. His reading thus constitutes multiple layers and fragments, which are repetitive, interlocked, incomplete, non-hierarchical, and in particular an immanent product of architecture space. Thus, both Markus's and Tschumi's readings of architecture project are an argument about space/place which considers the constituents of architecture space between people and place, and the interrelations, people-place, meaning, and event (Figure 6.8). Markus's reading projects the immanent social meaning and relations in space and Tschumi's projects the paradox between people and space immanent in architecture space. However, this argument also reflects many consistencies and unexpected similarities as will be explored in the following section.

Markus				
Conjunction	*Cause-and-effect **relation***	*Reading/ transcendent*	*Function form and space*	*Experience*
Interrelations ⇨	*People-place* ⇨	*Meaning* ⇨	*Arch. Space* ⇨	*Event*
Disjunction	***Paradox** double oscillation*	*~~No sense/~~ ~~No meaning~~*	*Transformation process and cross- programming* ⇨	*Meaning production/ immanent*
Tschumi				

◄ **FIGURE 6.8**
Architecture space in between Markus and Tschumi

Interrelations

Markus's and Tschumi's interrelations lie in between conjunction and disjunction respectively. However, these conjunctions and disjunctions should not be taken as in opposition. They both imply the independence of architecture space constituents; however, they also imply a different set of relations between them. These relations also work on two different levels.

Conjunction, as defined by Markus is a social product that considers classification and order of the constituents of architecture space through a cause-and-effect, hierarchal, linear, and structural relationship, represented through typologies and design guidelines; see Figure 6.9.

Connection both/and **Conjunction** ⇨ Cause-and-effect Outside/social	Hierarchal structure **Order and classification** ⇨ Well-defined similarities/ differences	Form Spatial structure Function **Building types and design guidelines (rules)** Boundaries and clusters Power relations

◄ **FIGURE 6.9**
Reading Markus's architecture space through conjunction

On the other hand, disjunction as defined by Tschumi, is an architectural process which considers the transformational sequence of space/place, which rejects a cause-and-effect, hierarchal, linear, and structural relation, hence, it considers a combination and permutation process of design; see Figure 6.10.

Accordingly, we follow Markus and Tschumi by using a constructed conversation in order to explore their readings of the interrelation of architecture space in between conjunction vs. disjunction, order and classification vs. sequences and transformations, urban typologies and guidelines vs. combinations and permutations.

Dis-connection either/or	Spatial, temporal, sensory, programmatic . . .	Breaks the binary of people/place into (space, movement, etc.)
Disjunction ⇨	**Sequences and transformations** ⇨	**Combinations and permutations**
Anti-synthesis Dislocation and juxtaposition Transgression	Reciprocal, indifferent, and conflict	Changes the pre-order of place relations

Conjunction vs. disjunction

Conjunction is an act of joining, a connection, combination, juxtaposition, and union of categories (Encarta-Online-Dictionary). It entails simultaneous occurrence through the connective logic of 'both/and', 'it is true if, and only if', all categories are true; i.e. it works through a cause-and-effect relation: when one category occurs the other follows (Blackburn, 2008).

Disjunction, on the other hand, is the act of disjoining, a dislocation, disconnection, and incoherence (Encarta-Online-Dictionary). It follows the dissociative logic of 'either/or' which does not necessitate a 'both/and' relation (Blackburn, 2008). Accordingly, disjunction does not refuse the both/and logic of conjunction. However, it rejects the cause-and-effect relation between the constituents of architecture space.

For Markus, conjunction entails the interdependency of the projected architecture space constituents, 'not at the level of the phenomena themselves' but at a deeper level that considers the social production of meaning (Markus, 1982a: 6). '[Conjunction] is not an internal relationship within the discourse 'architecture' but an external link in society'. It provides a typical interdependency of architecture/urban space, which is 'powerful, more appropriate . . ., and more dominant' (Markus, 1987: 484).

Such conjunctions could arguably, be called building types and, further, classification could, arguably, be the device which is the basis for the origin and development of building types. (Markus, 1987: 484)

For Tschumi, disjunction is 'a systematic and theoretical tool for the making of architecture' (Tschumi, 1987b: 213). It represents the relations between the mutually exclusive constituents. Through disjunction, these constituents are 'ultimately independent' (Tschumi, 1994b: xxi), they do not intersect, they affect each other when, and only when they interact (Tschumi, 1987b): 'an architectural element only functions by colliding with a programmatic element, with the movement of bodies, or whatever' (Tschumi, 2001: 213). The cause-and-effect relation between them is displaced through the disjunctive dissociative logic, to propose a new set of relations, dynamic process that embraces the inherit contradictions in their relation. Accordingly, Tschumi rejects the 'transparent' production of meaning through a cause-and-effect relation.

The concept of disjunction is incompatible with a static, autonomous, structural view of architecture. But it is not anti-autonomy or anti-structure.
(Tschumi, 1987b: 212–213)

Consequently, disjunction pushes architecture to 'interrupt its limits'/boundaries; the dynamic process of disjunction deconstructs architecture presence through dissociation between space and time (Tschumi, 1987b; Tschumi, 1994b). Disjunction 'implies constant mechanical operations that systematically produce dissociation in space and time' (Tschumi, 1987b: 213).

Accordingly, introduced strategies of disjunctions through:

- dissociation that replaces synthesis
- superposition or juxtaposition that replaces 'the traditional opposition' between space and function
- and finally, emphasises the dynamic process of dissociation, superposition, and combination that extends the architecture limits.
(Tschumi, 1987b)

Order and classification vs. sequences and transformation

In between conjunction and disjunctions, Markus and Tschumi read architecture space through order and classification on the one hand, and articulation (sequences) and transformations on the other. And they hence approach architecture space and design through typologies and design guidelines, and combinations and permutations, respectively.

Order and classification considers the arrangement of entities and categories through a hierarchical relation that considers their value and importance; accordingly they are classified into groups according to their types (Encarta-Online-Dictionary).

While articulation and/or sequence considers the connection and arrangement of these entities and categories (Encarta-Online-Dictionary), i.e. an articulation does not entitle hierarchical relations like an order, it only implies the connection that is yet to be described. And transformation considers the process of change of these entities (Encarta-Online-Dictionary). Accordingly, this implies a dynamic process that involves the non-hierarchical arrangement of the constituents of architecture space which simultaneously articulates and transforms between them.

Using categories and then arranging the categories into a systematic order is classification.
(Markus and Cameron, 2002: 43)

Markus considers classification as 'a function of language' for the study of architectural types and characters, one which is socially constructed rather than conceptual (Markus, 1987; Markus and Cameron, 2002: 16). Classification constructs a structural system that involves the identification of similarities and differences into categories, which are then distributed hierarchically, similar classes at the centre and dissimilar scattered at the margins (Markus, 1987; Markus, 1993; Markus and Cameron, 2002). This distribution is manifested through clusters and boundaries, which are simultaneously materialised through architecture design as building types and design guidelines; 'designing buildings is to subdivide and categorize spaces, their uses and their users' (Markus and Cameron, 2002: 16). Accordingly, Markus considers five steps of the classification process: a general reading, identifying the categories in this reading, relating these categories to space, designing spaces according to the constructed relations, and finally identifying rules and patterns that helps to control these constituents as people, place as well as people-place (Markus and Cameron, 2002).

These orders and classifications are demonstrated through architecture space [clusters and boundaries] both explicitly – categories of function – or implicitly – categories of users (Markus, 1987: 467–468). It is therefore necessary to re-emphasise Markus's interest in the silent points in these orders and classification, what is missing between clusters and boundaries, as well as the functional categories themselves.

> In any classification, there are empty sets – that is, elements or classes that could exist but about which there is silence.
> (Markus, 1987: 475)

> Any architectural sequence includes or implies at least three relations. First an internal relation, which deals with the method of work; Then two external relations, one dealing with the juxtaposition of actual spaces, the other with program (occurrences or events).
> (Tschumi, 1983: 153)

> Tschumi introduces two types of architectural sequences, internal and external. The internal sequence considers the architectural production process. However, this research is particularly interested in the external sequences, which considers the reading of architecture space through sequences of place, people and people-place, event, etc. These sequential readings of constituents of architecture space blur their definition through the multiplicity of the sequence rather than the unified well-defined classification.

> These architectural sequences read through the spatial (space/place), temporal, sensory (people), programmatic (context) etc., i.e. these sequences are projected through the primary framework of urban space. The spatial sequence is considered 'constant through our [architecture] history', which addresses space typology, the geometric transformations of space and form. The temporal considers these transformations through time. The sensory addresses

sequences of movement and perception in space,
which 'can be objectively mapped and formalized'.
Contextual sequence considers programmatic 'social
and symbolic' relationships (Tschumi, 1983: 153,
162, 154). Simultaneously, the idea of sequence could
be applied to all constituents of architecture space.

These sequences could be categorised as close, as
with 'people and people', which implies a closed
circuit of transformational process, or open, as
with people-place relationships, which implies
continuity of transformations in between 'people'
and 'place' as a constituent from one sequence
is added to a constituent of another, and so on.
'Alternatively, of course, architectural sequences can
also be made strategically disjunctive' (Tschumi,
1983: 168). Consequently, these sequences
could be 'contracted' – a continuous instant
transformation sequence between constituents
of place – 'expanded' – a 'gap' exists between
constituents transformations that 'becomes a space
of its own' – or 'combined of both expanded and
contracted sequences' (Tschumi, 1983: 165).

Markus (1982a) identifies the classification orders as
follows:

spatial (formal) order, which involves composition,
style and geometry; 'place' functional order, which
considers user patterns; 'people' structural order,
which considers 'place-people', 'form-function'
represented through building types nature/
built order, which follows Vitruvius's approach
to nature, 'the peaceful juxtaposition of the built
order and nature'; 'context'.
(Markus, 1982a: 8)

The deconstruction of Markus's text lies within his text. His empirical study
of architecture design, classification hierarchy, and typology did not 'uni-
formly mimic a social one' (Markus and Cameron, 2002: 74). Furthermore,
the expectedly tree-like 'hierarchal structures are not necessarily tree-like'
(Markus, 1987: 467). The boundaries between the classification orders are
not clear and well-defined, 'an element can belong to more than one class'.
Markus also, identifies the 'philosophical debate' developed about 'arbitrary'
relations of these orders and classification, which could conversely 'represent
natural real structures' identified through 'empirical observations' (Markus,
1987: 467). Accordingly, he recognises the mess of social reality as represented
through order and classification as well as the inability of a well-defined

hierarchical structure to approach this mess, the presence of empty sets as well as the replicating presence of some elements in these sets. The approach has to be more flexible and more dynamic and the boundaries less defined; the tree-like structure is replaced by a 'lattice'[3] (Markus, 1987: 467).

[Design guidelines and regulations] . . . in recommending what architects should or should not do . . . intend to constrain current and future building design . . . and they in turn design society.
(Markus and Cameron, 2002: 40)

Markus (1987) argues that conjunction relations find representation through types of spaces and buildings. Hence they assist the development of regulations and design guidelines and are considered as a manifestation of the architect's professional authoritative knowledge (Markus and Cameron, 2002). Conjunction helped the production of two discourses in the order and classification of 'place-people' relationship, namely 'form-function' through design guidelines and building regulations. These discourses gave a detailed description and prescription for form and space classification in relation to physical setting with particular interest in accommodating and expressing the space (Markus, 1987; Markus and Cameron, 2002).

[A] transformation process takes place within the functional discourse.
(Markus, 1982a: 5)

Tschumi (1983b, 1984) rejects the cause-and-effect relation between people and place sequences. As an alternative he introduces three sets of transformational processes: indifference, reciprocity, and conflict. Indifference entitles the total independence of architecture sequences, reciprocity entitles their total dependence, and conflict occurs where they contradict each other and 'constantly transgress the other's internal logic'
(Tschumi, 1983: 160).

'Part of a complex of transformational relations' is combinations. Fragments of the architectural sequences are 'recombined through a series of permutations' which implies changing the pre-order of the elements of each sequence

(Tschumi, 1984: 181, 180). Simultaneously,
transformational relations go beyond the
constituents defined as 'people and place', 'function
and form', etc. Elements of these constituents, like
'movement, space, event' are caught up in a set of
combinations and permutations.

The conflict . . . is deeper than that between
beauty and utility, or form and function. If all
building involves order-making, articulation,
division (between functions, between outside
and inside, between the space of strangers and
that of inhabitants, between nature and human
creation), then all of it is, or risks being, an
instrument of alienation or even imprisonment.
(Markus and Cameron, 2002: 24)

People-place

The dominant history of architecture, which is
the history of the signified, has to be revised, at
a time when there is no longer normative rule, a
cause-and-effect relationship between a form and
a function, between a signifier and its signified:
only a deregulation of meaning.
(Tschumi, 1989: 222)

Tschumi and Derrida . . . merely demonstrate their
inability to break the bonds of the most restrictive
of all constraints [cause-and-effect] on modern
design and criticism. . . . Deconstruction will not
read . . . everyday experience as meaningful.
(Markus, 1988: 343, 348)

A controversy concerning the projection of people-place developed between
Markus and Tschumi. On the one hand, Markus considers the production
of meaning through the traditional architecture cause-and-effect relationship
between people and place. However, the production of meaning transcends
to the architecture space. Meaning is developed by 'people', and is perceived
through function in a minimalist approach, e.g. form follows function – or
through a social production as in a maximalist approach. Tschumi, on the
other hand considers the paradox between people and place, which rejects
cause-and-effect relationships. As discussed, Tschumi approaches the paradox through the insertion of a third term, the pleasure-of-violence between
place and people experience. Thus, meaning is not a by-product of a cause-
and-effect relationship, but an immanent product of transformational process and programming of architecture space. It is viewed as an event outside
the reading of architecture space; however, it is presented in this section in

relation to Markus's approach. This controversy of people-place projection is deepened by Tschumi's attack on cause-and-effect relationships as well as by the transcendence of meaning and Markus's complementary attack on the deconstruction approach of Tschumi and Derrida: 'its practitioners add confusion; its theoreticians add amnesia' (Markus, 1988: 348). Accordingly, this section will approach the controversy of production of meaning through a cause-and-effect relationship and/or through the rejection of this relationship.

Cause-and-effect vs. paradox

The hierarchical cause-and-effect relationship . . .
is one of the great certainties of architectural
thinking . . . reassuring . . . that we live in houses
designed to answer our needs.
(Tschumi, 2001: 255)

The hierarchical cause-and-effect
relationships . . . go against both the real
pleasure of architecture . . . and the reality of
contemporary urban life.
(Tschumi, 2001: 255)

Markus reads architecture space through a cause-and-effect relationship between people and place, social context, and architecture space, which implies comfort, harmony, and homogeneity between them and represented through typologies and guidelines. He also recognises the inherit conflict between them. However, he re-draws on the Vitruvian tradition to address the conflict, which helps to increase the conflict rather than resolve.

The conflict between beauty and utility, or form
and function, is deep, and in design practice,
never fully resolved. . . . The more agonizingly
beautiful the created object, the more acute the
conflict.
(Markus and Cameron, 2002: 23–24)

[A]rchitecture must be conceived, erected and
burnt in vain.
(Tschumi, 1974 in Tschumi, 2001: 262)

On the other hand, Tschumi's rejection of the
dominance of 'cause-and-effect' is a rejection of:

• meaning as a by-product of this
relation meaning is a future construction through
the interaction between the different constituents

- synthesis relations between people and place
  - the supremacy of linear direct relations
  - the asymmetry between people and place where one constituent dominates the other, people and place interact and affect each other equivalently. It is 'impossible to determine which one initiates and which one responds' (Tschumi, 1981b: 122, 127).

He reads architecture space as a paradox between people and place as previously discussed, and inserts a third term 'pleasure' between them to subvert their duality and reject the cause-and-effect relation. Simultaneously, he read the violence between people and place, particularly between material space and body, as the body violates the materiality of space and vice versa (Tschumi, 1981b).

Why has architecture theory regularly refused to acknowledge such pleasures [violence] and always claimed . . . that architecture should be pleasing to the eye, as well as comfortable to the body?
(Tschumi, 1981b: 125)

Meaning vs. no-meaning

Strictly speaking, semiotics and structuralism propose language not as a metaphor for architecture but rather that architecture is a language.
(Forty, 2000: 80)

Meaning remains a controversy in architecture. The conception of architecture space in relation to language and semiotics lies at the heart of the controversy of reading meaning. A semiotic reading of architecture space emphasises form and physical space as the signifiers that produces meaning – the signified – through cause-and-effect relationship. Interestingly, both Markus and Tschumi read meaning beyond semiotics and signification. Markus and Cameron (2002) on the one hand, reads architecture as a language of analogies and metaphors; meaning is a social construction, a reading of meaning embodied in architecture space. Tschumi (2001) on the other hand, rejects the reduction of architecture space to language. Meaning is neither a prior – transcendent – construction nor is it immanent in the physical space; it is an event, which takes place in the future through the interaction between people and place. It is worth noting that Tschumi's rejection of the production of meaning through a cause-and-effect relationship is a rejection of the dominance of this relationship rather than its denial. At the same time, as demonstrated in the last section, 'cause-and-effect' is displaced by a transformational process that involves relations of indifference and conflict besides a reciprocity that echoes 'cause-and-effect'. In this sequence, Markus's and Tschumi's reading of meaning could be approached as two analogous perspectives rather than as contradictory.

[A]rchitecture is a language.
(Markus and Cameron, 2002: 1)

[A]rchitecture is not a language.
(Tschumi in Walker, 2006: 61)

Markus explores the strong relation between language and 'almost everything they [architects] do' (Markus and Cameron, 2002: 1). However, his approach goes beyond semiotics and structuralism. Semiotics approaches the linguistic meaning of words as an internal process detached from its context. Markus considers a 'pragmatic and socio-linguistic' approach to meaning and language, one which attaches meaning to 'the whole context: social, temporal, and spatial' (Markus and Cameron, 2002: 10).

To dismantle meaning showing that it is never transparent, but socially produced.
(Tschumi, 1987a: 201)

Groat (1981) has demonstrated how . . . social studies of place . . . have emphasised meaning, the signified, over the physical form the signifier. And architecture have tended to focus on the physical place rather than the meaning (Groat and Després, 1991; Chapter 5, this volume).

Buildings [architecture space] are primarily social objects. They carry meanings for society in general, and occupants and users in particular, which relate to asymmetries of power.
(Markus, 1987: 467)

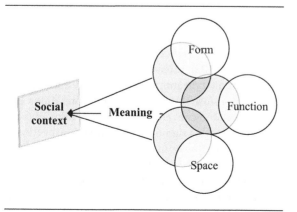

◀ **FIGURE 6.11**
Meaning in Markus's reading of architecture space

Through his reading of architecture space, Markus (1987) proposes a framework to study how architecture can 'carry meaning' to people (Figure 6.11). Form, space, and function are the three architectural discourses which primarily 'yield three sets of independent characteristics' (Markus, 1987: 467). Meaning is embedded within each and every set, as a relation between architectural space and people. These relations do not perform on an abstract architectural level, but are incorporated within the wider social context (Markus, 1982a). Accordingly, these sets interact on the level of meaning (Markus, 1982a). 'A focus on social relationships and the analysis of social structure appear to provide the links between the analysis and aid the discovery of meaning' (Markus, 1982a: 6).

J. Wolfreys:	What is the most widely held misconception about you and your work?
Derrida:	That I'm a sceptical nihilist who doesn't believe in anything, who thinks nothing has meaning, and text has no meaning. That is stupid and utterly wrong, and only people who haven't read me say this. (Wolfreys, 2007: 2)

Sign

Signifier → signified
Form → meaning
Form → function
Space → action

[T]he relation between form and meaning is never one between signifier and signified. Architectural relations are never semantic, syntactic, or formal, in the sense of formal logic.
(Tschumi, 1984: 185)

The reading of Tschumi's meaning was developed through a number of mixed manifestations. A rejection of meaning: there is 'no sense/no meaning' (Tschumi, 1987a: 200). The aim is to achieve an architecture which 'is a pure trace or play of language', 'an architecture of the signifier rather than the signified' (Tschumi, 1987a: 203). Interestingly, his reading does not reject meaning in architecture space, but it does refuse the established interpretations and associated conventions, i.e. the reading of form and meaning as signifier and signified respectively. Alternatively, meaning is a by-product of the architectural transformational processes. However, Markus (1988) associates Tschumi's rejection of meaning in architecture space to the approach to language of architecture through image and styles of himself and Derrida.

S E M. . . . If there was anything that could be seen as 'meaning', it was in the meeting of Space/ Event and Movement . . . a semantic coincidence.
(Tschumi in Walker, 2006: 42)

For Tschumi, 'S E M' is an 'ironic statement' that brings about the triangle of 'signification' and simultaneously rejects the 'embodiment' of meaning in architecture (Walker, 2006: 42). At the same time, meaning is the by-product of this sequential process SEM, as people interact and use place. Meaning is not an attribute of either people or space. It is not 'immanent in architectural structures and forms' (Tschumi, 1987a: 200).

Derrida, as discussed in Chapter 2, discards the signifier – represented by the physical form in architecture. Accordingly, the sign refers to an absent signifier, to a signified which refers to reality. Furthermore, a sign refers to another signified – for example classic styles in architecture. Accordingly, meaning is produced 'only by referring to another past or future element in an economy of traces' (Derrida, 2004 [1979]: 29).

Reality A←Signified A←Signifier←**Sign**→Signifier→Signified B→Reality B

The excess of meaning lacks meaning. But how can meaning be produced when signs only refer to other signs; when they do not signify, but only substitute?
(Tschumi, 1984: 176)

At the same time, meaning does not transcend architecture space through society, history, culture, and so on, where an architectural sign would signify another sign, historical for example. Tschumi hence seeks 'an architecture without any prior signification' (Walker, 2006: 60), which implies an architecture that does not draw on an external meaning existing in the urban context, social, history, culture, and so on. Accordingly, it 'frees' architecture from traditionally established conventions of meaning so that 'in future, [it] will be able to receive new meanings' (Tschumi, 1984: 174).

The reading of architecture through a transcendent meaning in history, society, and so on, imprisoned architecture space in a triangle of signification that subverted the physical space (form) and detached architecture from meaning.

History←Meaning←Form←Sign←**Architecture**→Sign→Form→Meaning →Society

> As an architect, one can encourage certain conditions for this use and misuse that will potentially entail a meaning, but one has absolutely no control over meaning. . . . [For example] if the context is gone, the meaning is gone.
> (Tschumi in Walker, 2006: 51)

Place: form/concept/concept-form

> [O]ur forefather only built their hut after they had conceived its image. This production of mind, this creation is what constitutes architecture.
> (Boullée, 1968 in Tschumi, 1975a: 34)

The discourse of place between Markus and Tschumi as discussed in Chapter 1, developed from a concern for a structured organisation of space and form into a resistance to this stability of structure and image.

Markus's reading emphasises the discourse on physical space. Like Canter (1977) and Relph (1976), he considers the physical settings and attributes of place. However, neither Canter nor Relph provide an illustration of the physical attributes of place. Markus provides a description of the experience of the physical space as form – style and composition – and space – urban morphology and typology (1982a, 1986, 1987) (see Figure 6.12).

▶ **FIGURE 6.12**
Markus's 'place'

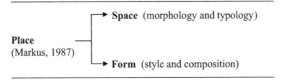

Place (Markus, 1987) → Space (morphology and typology) / Form (style and composition)

> Tschumi on the other hand is not interested in physical space, 'over the past years, there is one word I have almost never used, except in order to attack it: form' (Tschumi, 2010). He reads place through a confrontation between 'spaces and actions' (Tschumi, 2001), which blurs the definition of place becoming

contaminated by another category, 'people', or 'action'.
Moreover, Tschumi introduces the 'concept-form' in
between physical and conceptual space, 'a concept
generating form, or a form generating concept'
(Tschumi, 2010) (see Figure 6.13).

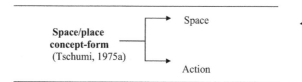

Space/place
concept-form
(Tschumi, 1975a)

Space

Action

Accordingly, this section will now explore place as form and concept. This
reading studies the boundaries and edges, space design, and movement-in-
space, and attempts to represent these through a sequential reading.

Boundaries and movement

Markus does not explicitly approach either the
boundaries and edges of a place nor the movement
within and between, but approaches their
architectural manifestation in place and through
architectural drawings, plans, sections, elevations
as well as 3D diagrams. Form represents 'the formal
properties of place and of the boundaries which
define it' (Markus, 1982a: 5), i.e. he considers the
architectural style, which involves the 'symbolic,
semiotic and abstract content of style' (Markus,
1982a: 5) and the architectural composition
which involves formal 'geometric properties, the
proportions, articulation, colour, ornamentation, and
surface treatment' (Markus, 1987: 468). Boundaries
are addressed through 'the number and location
of entrances from the outside'. And movement
is addressed through the architectural paths and
routes; 'the number of different routes to a space
(that is whether it is on one or more rings or part of
branching, tree-structure)' (Markus, 1987: 469).

Social studies of place, on the other hand, approach the architectural style
through the people: 'people's recognition or formal categorization of defined
architectural styles; people's stylistic preferences for a given building type;
and the social meaning associated with housing of particular styles' (Groat
and Després, 1991: 33). However, the study of composition is 'relatively rare'
in social studies, which consider either the complexity of the architectural
composition, or the 'significance' of certain physical attributes in relation to
people (Groat and Després, 1991), and similarly in relation to psychology and
behavioural process (Canter, 1977a).

The walls around the city have disappeared
and, with them, the rules that have made the
distinction between inside and outside.
(Tschumi, 1989: 216)

Architecture space, which is intrinsically related to
presence and materiality in physical space, is also
an abstract idea, a conceptual space. However, this
conceptual space shows a dual contradiction in
architecture thinking, which Tschumi considers as an
intrinsic attribute within architecture. This conflict
or contradiction goes back to a definition of 'space'
which relates space settings, place-making and the
setting of its limits and boundaries to architecture,
and the study of the 'nature' of space to philosophy
and mathematics, as is discussed in Chapter 1.
However, architecture has shown a genuine interest
in ideology and philosophy of space rather than
simply regarding the construction of physical space
'as the sole and inevitable aim of their activity'; 'one
must conceive in order to make' (Tschumi, 1975a:
33, 34). Simultaneously, the transformational process
of space involves the development of a sequence of
spaces through juxtaposition, addition, repetition,
etc. However, the spatial sequence goes beyond the
'formal composition', as it interacts with the sensory
space (1975a; Tschumi, 1987b: 208). The sequence
entails continuity and discontinuity between spaces,
which formalises the sequence body of movement-
in-space as well as establishing stand points (1975a;
Tschumi, 1983).

Typology and sequence

At the same time, space, the spatial experience
consists of morphology and typology. Markus
explores the formal properties of space,
although architectural style and composition
are not addressed on this level (Markus, 1982a).
Architectural theory in general examines the
'descriptive analysis' of the morphological properties
of place (Groat and Després, 1991).

On the other hand, Groat and Després (1991) consider typology, the formal and
functional structure of space as representing a 'significant point of intersection'
between architectural theory and social studies. However, social studies
approaches the functional structure, i.e. hospitals, and architectural theory has
approached the formal, i.e. L-shaped forms (Groat and Després, 1991).

Markus considers space typology through the depth, sequence, and permeability of the different spaces, i.e. the number of spaces and the way people pass through from the outside to an inside location (1982a, 1987).

Environmental-behaviour studies considered building types in relation to

- historical development, meaning embedded in the structural properties of the form
- socio-cultural meaning linked to certain typologies, as well as
- social (people's) cognitive representation of these typologies.

(Groat and Després, 1991).

Alternatively, Markus considers the 'abstract morphological systems' (Markus, 1982a: 5) 'the structure of space, sequence and linkage' between the different spaces (Markus, 1987: 469).

Conversely, social studies rarely approached the descriptive analysis of space morphology; rather they tended to understand through their relation with the wider socio-historical context (Groat and Després, 1991).

Tschumi reads space typology and morphology through sequential transformations of 'geometric forms' 'constructed, step-by-step – or deconstructed – according to any rule or device' (1983b: 156).

A sequential reading

Markus, in a similar way to Groat and Després (1991), reads place through style, composition, urban morphology, and typology, which is associated to physical space. Simultaneously, Groat (1981), as discussed in the previous chapter, criticises the model of place developed by Canter (1977a) for its exclusion of architecture space, particularly the physical, and she proposed the inclusion of this set of place. Interestingly, Canter (1997), in the discourse of the facet theory of place, includes Markus's reading of place, which also involves function, as a representation of the physical attributes of architecture space.

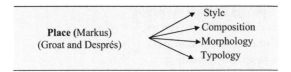

Tschumi on the other hand, perceives place through a sequence of concept (idea-to-space), physical (space-to-space), social (body-to-space), contextual (context-to-space), virtual (electronic space-to-space), etc. This reading appears to engulf Markus's concept of place through a sequence of physical spaces (space-to-space), which involves the transformational process of space geometry, organisation, form, and boundaries. However, Tschumi's reading also involves the contamination of these sequences through their interaction with other constituents. For example, the spatial sequence involves the transformation of spaces, paths, and links, and this helps to formalise the body movement-in-space through a sensory sequence.

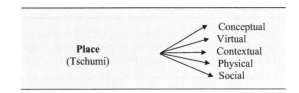

In Figure 6.14, I represent a sequential reading of urban space which interacts with both people and context. The reading implies non-hierarchal permutations of the constituents of place, people, and context. Simultaneously, the boundaries of these sets as well as their constituents are not well defined

▶ **FIGURE 6.14**
Urban space: a
sequential reading

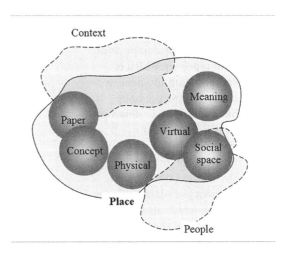

despite their mutual exclusiveness. Place involves physical space, conceptual space, paper (architectural drawings, etc.), social space etc. Place is considered an abstraction and/or generalisation of each of these categories which would be addressed sequentially in context. Place and people interact through social space. However, as previously explained, both sets are repeatedly contaminated through body movement, perception, etc. Finally, it should be noted that meaning is dislocated by the split between author/reader, designer/user. While the author/designer reads a meaning through people-place interaction, meaning escapes the sequence to inhabit architecture space through user interaction without prior planning or control.

People: function/programme/event

> The new questioning of that part of architecture called 'program', or 'function', or 'use', or 'events', is fundamental today . . . in our contemporary society, programs are by definition unstable.
>
> (Tschumi, 2001: 20–21)

This reading of 'people' extends Markus's and Tschumi's development of Vitruvius's 'Utilitas' as discussed in Chapter 1. Markus continues Vitruvius's tradition through his discourse on function, which like Utilitas considers the space-of-function rather than the function-in-space. Tschumi on the other hand, is concerned with body activities rather than function. Accordingly, he considers the displacement of the space-of-function brought about by the intrusion of the event into architecture space. He also emphasises the movement of the body-in-space which generates space-of-movement.

Facet A, in Canter's (1997) facet theory of place, considers the differentiation between spaces-of-function through typology of place, [particularly] central and peripheral [spaces-of-activities].
(Chapter 5, this volume)

The reading of 'people', in this section, is particularly interesting as it highlights many conflicts and contradictions between Markus and Tschumi, but also demonstrates consensus as they appear to echo each other in different aspects. This apparent agreement helps to bring together both readings as well as extend the reading of 'people' from a concern with function to a concern with body movement articulated through programme and event. This reading therefore explores people as an extension from body-in-space to context, from function to physical space, from programme and event and finally presents a sequential reading of 'people' constituents of place.

Body-context (social space)

Body-context is a projection of the twentieth-century reading of architecture space in which Markus (1982a) focuses on the social production of architecture space, in an apparent similarity to the 'maximalist' approach.

Tschumi (1980), on the other hand, rejects the separation between the two approaches and considers their integration as discussed in Chapter 1.

> '[M]inimalist' . . . concentrates on the details in architecture, style technique, etc. The maximalist, on the other hand, extends to the urban context, social, political, as well as programme.
> (Chapter 1, this volume)

In 'The Cultural Park for Children', the community (society) was involved in the verbal discourse on the park. Accordingly, at their request the park plan was modified to include a library for children.

Markus considers the development of the discourse on 'people' through a cause-and-effect relationship to 'place'. '[F]ormal and spatial solutions . . . embody the functional statement' (Markus, 1982a: 5). Function statements are produced through social 'verbal' discourse and consequently involved in the design process. Markus (1988: 343) also considers the development of architecture briefs and design guidelines to accommodate the 'body' needs in space. The 'body' here refers to user needs rather than the materiality of the body as understood by Tschumi; see Figure 6.15.

▶ **FIGURE 6.15**
Markus's reading:
context-body-space

Social context (*verbal*) ⇨ Function
Design guidelines and/or program
(*Body/user needs*)
(*cause-and-effect*)
↓
Architecture space (*Formal and spatial solutions*)

[S]uddenly, the body was there, regardless of what one thinks, whether one likes it or not.
(Tschumi in Walker, 2006: 27)

The Cultural Park for Children
In between
Community/child
Is a reading between
Social context/
body-in-space
Where the reading of social context
is dominant

Tschumi (2001: 3) emphasises the inclusion of the physical 'movement of bodies in space' along with social actions and events within the urban context. This inclusion is a by-product of violent relations between the materiality of the body and physical space (Tschumi, 1975a). This helps the 'articulation' between the body-in-space and social context, as discussed in Chapter 2. Accordingly the reading of people recognises and extends from inside the physical architectural space to the wider socio-cultural context, Figure 6.16.

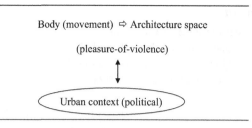

◀ **FIGURE 6.16**
Tschumi's insertion
of body (movement)
in space and
context

Function-place (physical space)

> Form now rules the day by speaking of function.
> (Markus and Cameron, 2002: 28)

For Markus, the relation between space (form)
and function 'is not in question – only . . . the
interpretation' (1993: 35). The cause-and-effect
relation simultaneously blurs the definition
between function and meaning. Therefore the
product is 'speaking architecture' that provides
'a direct understanding of a building's function'
(Markus and Cameron, 2002: 28). This relationship
operates in either a minimalist – body – or
maximalist – context. The first is interested in the
'function experience' of place which is concerned
with the architecture programme and place
typology (Markus, 1987: 468).

This relationship implies that each function
has 'predictable formal consequences' through
space morphology and organisation as well as
location and furniture (Markus, 1993: 37). The
maximalist approach, on the other hand, is
interested in the social function of space, function
as implying a symbol of meaning 'speaking
through allegorical and metaphorical forms'
(Markus, 1993: 34; Markus and Cameron, 2002).
Markus and Cameron (2002) also highlights the
attachment of the latter approach to classical
architecture styles.

The reading of Cairo space through
a monolithic representation of the
historic architecture style of the old
Islamic city, is a another emphasis
on the social context of the city – a
maximalist approach – rather
than the minimalist experience
of Cairene – the body – in Cairo
space.

Tschumi's reading as he himself explains, aims
to 'reinstate' people 'function' to the architecture
discourse, in opposition to the dominance of
physical place 'form' (Tschumi, 2001: 3–4). This
is evident in his emphasis on the experience of
body-in-space, and programme and event as will be
discussed in the next section.

Program-event

For Markus, the programme is the social construction of space function through 'verbal descriptions' developed into architectural briefs by architects and professionals (Markus, 1982a).

> Any predetermined sequence of events can always be turned into a program.
>
> (Tschumi, 1983: 157)

> Tschumi also reads the architecture programme as a descriptive brief of the different involved spaces-of-function; see Figure 6.17. However, he does not consider the inclusion of people, and/ or society in the process. In opposition, the event is 'a turning point – not an origin or an end' (Tschumi, 2001: 256). The event happens; it is neither organised nor arranged. Consequently, Tschumi (1981b) introduces the concept of 'programmatic sequence', which encompasses the use of space, events, etc. This is considered to be a particularly violent sequence as it interacts with the physical sequence.

▶ **FIGURE 6.17**
Tschumi's program-event

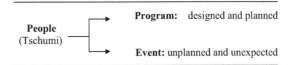

A sequential reading

> 'Program' is not the same as 'use', 'use' is not the same as 'function', 'function' is not the same as 'event'.
>
> (Tschumi in Walker, 2006: 27)

In a similar way to the reading to place, Figure 6.18 shows people as a sequence of function, use, programme, movement, etc. At the same time, the concept of 'people' is considered an abstraction and/or generalisation of each of these categories, and would be addressed in a sequence in urban space and context. Boundaries and hierarchy between these categories are not well-defined despite their mutual exclusiveness. People interact with place through the physical space that simultaneously extends to reach the context. Meaning is again dislocated from the sequence; meaning is a category of both the sequences of people and place, but meaning does not belong to sequence. Finally, the event acts as the supplement that escapes the boundaries of the sequence.

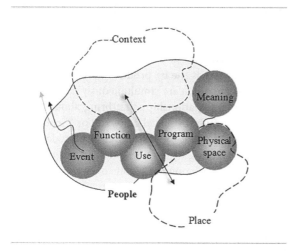

◀ **FIGURE 6.18**
People: a sequential reading

ON THE MARGIN

The review of architecture space in this chapter, between Markus and Tschumi, is a reading between the continuity and discontinuity of the traditions of architecture classicism as discussed in Chapter 1; and the paradigmatic shift in the reading of Khōra, explored in Chapter 3, between the traditional reading of the 'well-defined' categories and linear relations, and the new reading of the 'not-well-defined' categories, which creates a complex network of relations negotiations through their continuous oscillations between each other. At the same time, the review of social space in the previous chapter associated the reading of Markus with the facet theory, and Tschumi with deconstruction. Both readings recognised the dynamics and complexities of urban space; however facet theory attempted to provide a structured framework to categories and classify this reading whereas deconstruction attempted to further destabilise this relation, 'to put architecture into crises' as endorsed by Tschumi. Interestingly, the review of the facet theory of place associated Markus's discourse on architecture space to the physical space and Tschumi's to the social.

Consequently, Markus attempted to read place in correlation with language and discourse analysis that perceived place content as vocabulary, technique, organisation and composition, style, etc. The different categories of place are thus well-defined as form, space, and function. However, the boundaries between them are blurred through the relationships between them. Finally, he emphasises the value and need to look for and recognise the silent points in each of these categories. Tschumi, on the other hand, reads architecture space in relation to other disciplines, for example philosophy, through the practices of intertextuality, transgression of limits and boundaries, and transposition of metaphors. Moreover, these practices reject a synthesised reading of the violently introduced fragments and layers to architecture space which becomes contaminated.

Interestingly, both Markus and Tschumi recognise the conflicts and multiplicities associated with architecture space. However, Markus attempts to control and regularise them through a 'well-defined' reading, whereas Tschumi attempts to emphasise and exaggerate them. Accordingly, for Markus, architecture space is the by-product of design guidelines and pre-defined place typologies, which are simultaneously fashioned through conjunctive cause-and-effect relationships that bring about the transcendent reading of meaning through the social context. For Tschumi, on the other hand, architecture space is the by-product of unprecedented transformations and cross-programming which rejects the transcendence of meaning through cause-and-effect relations in favour of the disjunctive paradox of architecture space and which varies between reciprocity, indifference, and conflict. Meaning is thus dislocated into the immanent reading of architecture space, to come through the unplanned event of the interaction between categories of people and place.

Accordingly, the anti-synthesis practice helps the construction of architecture sequences, reading instances of categories of place, for example 'people', which recognise the multiple layers and fragments of each category which both extends to other constituents, e.g. context, and involves fragments of other sequences, e.g. physical space.

NOTES

1 This was first brought up as 'a reader complained that Tschumi had failed to cite Thomas Kuhn's (1970) book "The structure of scientific revolutions", although he had . . . almost integrally copied a passage from it . . . simply replaced the word "science" . . . with the word "architecture"' (Martin, 1990: 29).
2 Term used by Martin (1990).
3 In a reference to the '*lattice*', consider the '*Rhizome*' introduced by Deleuze and Guattari (1983) in *On the Line*; see also *A Thousand Plateaus* (Deleuze and Guattari, 1987).

TO BE CONTINUED . . .

In this interlude, I reflect on the book's journey to explore the reading of urban space; the singularity of this reading space, highlighting the perspective of various authors/readers, and the implication of reading space as dynamic relations rather than social, physical . . . spaces and attempting to relate them. This discussion thus constitutes two main points. The first discusses the reading of a complex, chaotic book subject, the mess of reading urban space both empirical and theoretical, particularly working within a post-structuralist background particularly deconstruction, between verification and recognition. The second point questions the reading of 'urban space' and considers the implications of the use of reflexivity to develop the primary framework of urban space. Accordingly, it proposes the 'reflexive reading of urban space' through various fragments presenting the sequences and constituents of urban space. However, this new reading of urban space, must remain without a conclusion to be continued . . .

THE MESS: VERIFICATION VS. RECOGNITION

> If this is an awful mess . . . then would anything less messy make a mess of describing it? . . . won't help us to understand mess.
>
> (Law, 2003: 1)

The book goes beyond the anticipation of a single coherent reading of reality to recognise the mess of its subject, the reading of urban space, both empirical and theoretical. It considers approaching the theoretical mess of urban space in order to approach the empirical mess of Cairo urban space. Urban space is hence read through a complex network of relations between the abstract concept and people's experience and everyday activities, while remaining embedded in the wider urban context: social, economic, historic, political, cultural. At the same time, theories of urban space are then developed through a multitransdisciplinary reading using philosophies of space/place interested in the exploration of its nature, architecture theory interested in the making of urban space, and social studies with its particular interest in the experience of people in urban space. The multitransdisciplinary practice enables a dissolution of the boundaries between these disciplines where they interact, producing many similarities, differences, and inconsistencies. The empirical study of Cairo space is another reflection of the complexity and chaotic nature of the study of urban space with a particular emphasis on its relation to time, people, and context.

Simultaneously, the approach to the demonstrated mess developed between a structuralist 'well-defined', singular reading with a hierarchical classification and categorisation of the book subject, similarities at the centre and dissimilarities on the margin, and a post-structuralist 'not-well-defined' reading that accepts ambiguity and plurality, and addresses equally both primary and secondary data, and the 'trivial and non-trivial' though showing a special interest in the marginalised data. A structuralist approach, on the one hand, understands urban space, in terms of separate elements and linear relationships; analysis attempts the isolation and logical interpretation of these through a synthesised formulation of a unified singular reading, i.e. the simplest hierarchical framework. A post-structuralist approach, on the other hand, understands urban space as a network of complex and negotiable categories, subcategories, and relationships, and recognises the multiplicities, diversities, and inconsistencies of their interpretations while favouring a singular or best reading. Furthermore, it attempts the continuous destabilisation of these interpretations and their authority. In summary, both approaches recognise the mess of the book subject, both theoretical and empirical. However, the first approach attempts the hierarchical classification of this mess, its similarities and differences, through the structural primary framework of interpretation, whereas the second attempts the destabilisation of this interpretation through continuous exploration and awareness of the mess itself. Structuralism therefore risks simplification and misrepresentations, while post-structuralism intrinsically runs the risk of a continuous, never-ending exploration. However, the development of the book approach between the two frames of reference helped to minimise these disadvantages. The following section reflects on the alteration of the aim in order to develop a structured 'framework' to projection of urban space, to non-structured 'reading strategies'.

A REFLEXIVE READING OF URBAN SPACE

> [I]t is . . . necessary to recall the paradoxical relationship between architecture as a product of mind, as a conceptual and dematerialized discipline, and architecture as the sensual experience of space and as a spatial praxis.
>
> (Tschumi, 1976: 66)

'A reflexive reading of urban space' . . . the threshold was: 'how to approach the reading of urban space?' This question was challenged by the multiplicities, dynamics, and complexities of both the phenomenon and the study of urban space. The latter drew on philosophy – primarily interested in exploring the nature of space/place – social studies – involved in the study of people's experience of place – and architecture theory – concerned with place-making – as well as the empirical vignette of Cairo urban space. The approach to this question thus evolved from a concern to develop a multi-disciplinary framework which operates on two levels: urban space and context, to consider a reflexive reading from among the multiple projections of urban space that extend from the concept to content to context.

At the same time, this book was itself developed through a reflexive reading. A primary framework of urban space was developed at an early stage through an initial reading of the theories of space/place involving: Canter (empirical based, 1977) and Relph (phenomenological, 1976), Markus and Tschumi (architectural, 1980s~), and Canter (facet theory of place, 1997). This framework recognised the general categories and relations and was introduced as the primary framework of interpretation. However, in order to develop this framework into reflexive reading strategies, these theories were re-approached as the study object, as theoretical case studies of urban space reading.

Author/reader

The multidisciplinary reading of urban space thus explores the role and perspective of the author/reader as designer, interpreter, and participant. The designer – architect, urban designer, and planner – is concerned with conceiving urban space through the imaginative experience of the first reader. The reading of architecture space has long been associated with Vitruvius's trilogy: Firmitas, Utilitas, and Venustas and has continued through the post-Vitruvian trilogy of space, function, and form. Tschumi (2001) has demonstrated the unperceived displacement of Vitruvius's readings in architecture today, through space, movement, and event. Furthermore, he complemented this displacement through another trilogy – concept, content, and context. The discourse on urban space, on place-making, also involves an institutional and administrative contribution whose role is particularly overemphasised in the context of Cairo space, outweighing the contribution of architects and urban designers.

Philosophers and social researchers are considered as interpreters, who are particularly interested in the exploration of the experience of people and place. Philosophers, on the one hand, are concerned with the study of the nature of space/place, container, relational, internal, external, extended, etc. These studies, as introduced in this book, involve three periods: the ancient which approached space/place as a whole, with the complication of subsuming chóra under Aristotle's topos; the modern which helped the division of this whole through a set of dualities and which empowered space and emphasised the phenomenological experience of the body; and finally, a return to Khôra which brought urban space to heterogeneity, indirection, and complexity. Social studies on the other hand, are more concerned with people's experience, their activities, behaviour, etc. as portrayed through empirical studies. Early social studies adopted a positivist perspective which approached the reading of a well-defined place presented as elements and

To be continued . . .

relations. The later structuralist approach recognised the messiness of the dynamics and complexities of place, but also attempted the classification and categorisation of this mess through a well-defined framework of reading.

Finally, there are the participants, that is, the users of space, community, passers-by, and so on, whose role is not empirically involved in the reading of urban space in this book. However, the participants' role is carefully approached as an integral part of this reading: the 'people' presented as a study of relationships, people's relationship to place – body movement, experience, event, etc. – and people's relationship to others – the self, the other and others and to personal, social, and cultural space. Simultaneously, Khōra, as introduced by Derrida, is a unit; not a homogenous unit, but a unit nevertheless. These multiple spaces, of architecture, philosophy, and society, do not present different urban spaces/places but different perspectives that co-exist even when in conflict. This reading could thus help bring them together, through their interaction and communication into a relationship which recognises and accepts integration, indifference, and conflict.

The Cultural Park for Children

The conflict between the architect's perception of the social space of community and the manager's perception of the child's user space could be brought into interrelation through the space – the park – that brings together the conflict.

By the way of Khōra

> [T]here is only one Khōra . . . however, divisible it be.
>
> (Derrida, 1995: 97)

A significant reflexive instance is introduced through the study of Khōra, who is involved in both the reading of urban space in architecture and philosophy and in the marginal reflections on the development of this reading. Khōra plays a fundamental role in the development of the new paradigm of space/place. The traditional autonomous elements of place: body, space, and mind, evolved from separatist linear relations into heterogeneous, complex negotiation networks. Khōra is both the abstraction and realisation of urban space, and vice versa, as concluded in Chapter 3. At the same time, Khōra rejects her submission to matter, a subset of topos, as well as the habitation of architecture space in topos. Thus, she extends with architecture space into the milieu: urban context (Figure TBC.1). Furthermore, Khōra is a relational space, continuously moving between two types of oscillation: neither/ nor and both/and, rejecting the determinism of well-defined boundaries and cause-and-effect relationships.

▶ **FIGURE TBC.1**
Khōra: and architectural and relational space

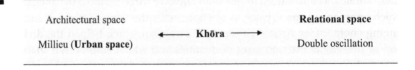

Architectural space		**Relational space**
	⟵ **Khōra** ⟶	
Millieu (**Urban space**)		Double oscillation

Khōra, being an intermediate space, is the reciprocal of temporal readings of peaceful, indifferent, and conflictual relationships between the double oscillations but without possessing any of their properties. At the same time, she is always on the move; as soon she is defined, she shifts. She presents the space in between concept/content, concept/context, and content/

context (Figure TBC.2). She also oscillates internally between the rational/
emotional concept, the body/space content, and the context.

> The separation of conception – a mental imaginative process . . . from
> realization which requires materials . . . to make drawing concrete, now
> seems obvious.
>
> (Markus and Cameron, 2002: 26)

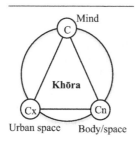

◄ **FIGURE TBC.2**
Khōra: a relational
space

The development of the concept of modern space initiated a separation
between space and place, space and matter, etc. which problematised their
reading and encouraged a return to ancient philosophy. Subsequent readings
emphasised the materiality of bodily experience in space, creating another
binary: body/space, which implied the reading of mind as separated from
body. Khōra, the contemporary re-readings of Plato's chóra, attempted the
deconstruction of these binaries and their consequent separatist relationships. These readings continued to emphasise the context, thus separating
context from content; the content – body and space – is located in context,
creating the binary reading of urban space – content/context, or a trilogy
of space, place, and context. Tschumi (2004) developed the trilogy: concept
(space), content (body and space), and context which he applied to architec-
ture/urban space. My proposition is that while this separation of the context
has further implications for the reading of space/place, at the same time, the
potential for going beyond this separation remains doubtful.

Urban space is a space of knowledge of place in context (Figure TBC.3). It
is concerned with place-making which extends in between the concept: the
reading of urban space, the process: design, drawings, modelling, and so on,
and the construction. Khōra, on the other hand, oscillates in between the
space of thinking and the space of doing, each of which oscillates between
the abstract idea: intelligible and inspirational, and the material: sensible
and pragmatic (Figure TBC.4). However, Khōra, the reciprocal, is not the

Space of knowledge	← **Urban space** →	Design – drawings –
Concept – **context**	Content	construction
of	(body and space)	process
	in <u>context</u>	

◄ **FIGURE TBC.3**
Urban space

To be continued . . .

▶ **FIGURE TBC.4**
Khōra

Space of thinking	← **Khōra** →	Inspiration – pragmatism
Intelligible – sensible	urban space content (in-between)	Space of doing

content of urban; she is the passage to the delivery of this urban space. At the same time, the urban context escapes the content to become the space of knowledge: the sensible; the context becomes the sensible space of knowledge of urban space in context. The urban context is thus immanent in urban space, where the transcendent space of knowledge is approached through a temporal frame of immanence (Figure TBC.5). Accordingly, the urban context reads multiple urban spaces: economic, social, architectural, political, and so on, which simultaneously and continuously displaces each other. This understanding of the relationships in urban space – rather than between content, concept, and context – both questions and destabilises the reading of urban space, which is further discussed in the following section.

▶ **FIGURE TBC.5**
Urban contextual space: sensible space of knowledge immanent in place content

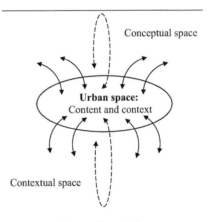

How can architects and urban designers contribute towards the reading and making of ~~Cairo~~ urban space?
The reading of place through conjunctions, cause-and-effect relationships, and so on, helped the production of design guidelines, laws, and legislations to control and regulate place-making, and provided 'well-defined' typologies of space/place in relationship to the function. This helped the production of several books and documents to guide and aid the contribution of architects and urban designers towards place, but which actually controlled rather than aided them.

The reflexive reading of fragments and sequences of architecture/urban space through disjunctions and double oscillations cannot intrinsically provide rules and regulations to control the production of urban space. However, this reading holds the potential for developing the awareness of architects and urban designers about the multiplicities and dynamics of urban space and for integrating theory with practice. This potential requires further study and examination, in close relation to both architecture design and education.

Reading fragments

The readings of Khōra, architecture, and social space have demonstrated the awareness and recognition of the dynamics and complexities within urban space, developing into multiple readings. Accordingly, ~~sets facets~~ sequences and ~~elements categories~~ constituents of urban space involve people: body, movement, use, function, programme, and so on; place: concept, drawing, physical, virtual, and so on; context: physical, social, economic, political, historic, etc.; and time: past, present, future – planned and unplanned (see Figure TBC.6 and Figure TBC.7).

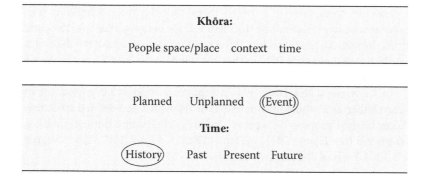

Khōra:

People space/place   context   time

◀ **FIGURE TBC.6**
Khōra: a reading sequence

Planned   Unplanned   (Event)

Time:

(History)   Past   Present   Future

◀ **FIGURE TBC.7**
A temporal sequence

The interrelationships between these sequences vary between inter-section, union, juxtaposition (peaceful and violent), conjunction, and disjunction. The temporality of a reading instance and multiplicities of the constituents allow these variations in between their relationships: for instance, two sequences can intersect through one instance and differ through another. The sequential reading rejects hierarchy, linearity, and well-defined structures. The relationships between the constituents help the construction of meaning, experience, and event through ~~cause-and-effect~~ double oscillations transformations and cross-programming, etc.

At the same time, the reading of urban space is an attribute of the perspective of the author/reader, presented earlier as designer, interpreter, and participant (Figure TBC.8). The practice of reading urban space is a sequential relation between mind and place – body and space. As discussed throughout this book, these relations are rational and/or emotional and are reflected through perception, cognition, and conception (Figure TBC.9).

Author/reader:

Designer – interpreter – participant

◀ **FIGURE TBC.8**
Author/reader sequence

Rational emotional

Mind/body

Perception cognition conception

◀ **FIGURE TBC.9**
Reading sequence

Urban space, like Khōra, is a unit, divisible, non-homogenous, multiple, complex, and dynamic but, nevertheless, a unit. However, these complexities have shown a persistent resistance to previous readings, which considered either an understanding of the whole phenomenon of urban space or a particular aspect of it. In general, these readings attempted to introduce frameworks and classifications in order to identify and structure the boundaries, the involved constituents – categorised as central and marginal – the silent points or gaps within the well-defined boundaries, and ultimately, the internal structure and relationships between these categories. However, these readings have revealed several difficulties and inconsistencies. Primarily, the boundaries of the constituents of urban space demonstrate a tendency to dissolve in between them, blurring their definitions and, more particularly, implicating the relationships between them. At the same time, the approaches to reading these relationships recognise the dynamics attributed to them even though they continued to be read through a linear framework.

In the view of this book, it is not possible to attempt to read urban space either as a whole phenomenon, or through a pre-defined structure. Accordingly, I propose to approach urban space reading through multiple fragments (see Figure TBC.10, Figure TBC.11, Figure TBC.12, and Figure TBC.13). Each of these fragments presents:

- A temporal instance which reads through a single perspective, context, space, and author. However, it also recognises and relates to other reading fragments.
- A singular reading, which approaches the reading of each fragment (for example people) from within, while simultaneously being influenced by the discourse (for example place), context (for example institutional space), and author (for example designer).
- A reading event, which while embedded in urban space and context, is yet-to-come unexpectedly and thus meaning is established afterwards.
- A paradox, which considers the deconstruction of binary representations of urban space through a third term which oscillates between them.

▶ **FIGURE TBC.10**
Context/space
reading between a
disciplinary and a
setting sequence

Architecture social philosophical urban

Context/space

Economic political temporal institutional social

▶ **FIGURE TBC.11**
Space/place reading
sequence

Container relational internal/external extended

Space/place

Conceptual physical virtual social

> Mind body society culture programme
>
> **People**
>
> Use activities function behaviour movement

◀ **FIGURE TBC.12**
People reading
sequence

> **Khõra**
>
> Experience event meaning

◀ **FIGURE TBC.13**
Meaning: reading
sequence

- A relational reading which approaches the relationships between the constituents of urban space rather than reading them as separate constituents and links. For example, people are read in relation to their economic context, rather than being studied separately as people and economic context and then considering the relationship between them.
- A reading sequence which approaches the paradoxical relationships as a transformational process without hierarchy or classification of structures and which thus provides multiple combinations and permutations.

Accordingly, these fragments should be approached through a dissociative reading. This dissociation allows their juxtaposition and superimposition without prior order or intended meaning. For example, the reading of place considers two sequences: first, fragments, in which one studies the nature of place: container, relational, etc., and second, which reads the attributed spaces: conceptual, material, social, and so on (Figure TBC.11). Another sequence reads people as fragments of uses, activities, functions, and so on, and mind, body, society, and so on (Figure TBC.12), and meaning is read through a sequence in between experience and event (Figure TBC.13).

The proposed fragmentation of the reading of place holds great potential for approaching the inherited complexities and dynamics within. At the same time, this reading recognises the exclusiveness of each fragment whilst allowing for the development of yet more readings. . . . This new reading of place, however, must remain without a final conclusion in a post-structural thesis. The fragments, outlined above, require further study, both individually, and in relation with each other.

To be continued . . . or re-read again. . . .

BIBLIOGRAPHY

Abaza, M. 2006. Egyptianizing the American Dream: Nasr City's Shopping Malls, Public Order, and the Privatized Military. *In:* Singerman, D. & Amar, P. (eds.) *Cairo Cosmopolitan: Politics, Culture, and Urban Space in the New Globalized Middle East.* Cairo, New York: The American University in Cairo Press.

Abdelhalim, A. I. 1978. The Building Ceremony. Unpublished Doctoral Thesis. Berkeley: University of California.

Abdelhalim, A. I. 1986. Thinking Paradise. *The Islamic Environmental Design Research Centre, Environmental Design: The Garden as a City,* 68.

Abdelhalim, A. I. 1988. A Ceremonial Approach to Community Building. *In:* Sevcenko, M. B. (ed.) Theories and Principles of Design in the Architecture of Islamic Societies. Harvard University and the Massachusetts Institute of Technology, Cambridge, Massachusetts: The Aga Khan Program for Islamic Architecture.

Abdelhalim, A. I. 1989. On Creativity, Imagination, and the Design Process. *In:* Serageldin, I. (ed.) *Space for Freedom: The Search for Architectural Excellence in Muslim Societies.* London: Butterworth Architecture.

Abdelhalim, A. I. 1996. Culture, Environment, and Sustainability: Theoretical Notes and Reflection on a Community Park Project in Cairo. *In:* Reilly, W. (ed.) *Sustainable Landscape Design in Arid Climates.* Geneva: Aga Khan Trust for Culture.

Abdelhalim, H. 1983. El-Houd El-Marsoud, Cairo. *In:* Khan, H. (ed.) *Mimar: Architecture in Development* (Vol. 8). Singapore: Concept Media Ltd.

Abdelwahab, M. A. 2009. *Reading Place: The Cultural Park for Children. Forum,* 9, 1–12.

Abdelwahab, M. A. 2010. *Cairo, Khōra and Deconstruction: Towards a Reflexive Reading of Urban Space.* PhD in Architecture, Newcastle University.

Abdelwahab, M. A. 2012. Towards a Relational Reading of Place. *In:* Durmaz, B. A. (ed.) *Designing Place International Urban Design Conference.* Nottingham, UK: University of Nottingham.

Abdelwahab, M. A. 2013. Cairo: A Deconstruction Reading of Space. *In:* Kellett, P. & Hérnandez-Garcia, J. (eds.) *The Contemporary City: International Research Perspectives.* Bogota: CEJA (Centro, Editorial Javeriano).

Abdelwahab, M. A. & Serag, Y. 2016. The Blessings of Non-Planning in Egypt. *In:* Albrechts, L., Balducci, A. & Hillier, P. J. (eds.) *Situated Practices of Strategic Planning: An International Perspective.* Abingdon, New York: Routledge, Taylor and Francis Ltd.

Adham, K. 2004. Cairo's Urban Déja Vu: Globalization and Urban Fantasies. *In:* Elsheshtawy, Y. (ed.) *Planning Middle Eastern Cities: An Urban Kaleidoscope in a Globalizing World.* London: Routledge.

Agamben, G. & Heller-Roazen, D. 1999. *Potentialities: Collected Essays in Philosophy.* Stanford, CA: Stanford University Press.

AKAA. *Aga Khan Award for Architecture* [Online]. Available: www.akdn.org/akaa.asp [Accessed 29 January 2009].

Akbar, J. 1992. Cultural Park for Children, Cairo, Egypt. *On-Site Review Report.* The Aga Khan Award for Architecture. Massachusetts: The Aga Khan Program for Islamic Architecture.

Aga Khan Trust for Culture [Online]. www.akdn.org/aktc.asp [Accessed 29 January 2009].

Al-Ahram-Weekly. 2000a. Derrida in Cairo. *Al-Ahram Weekly*, 10–16 February 2000, p. 468. http://weekly.ahram.org.eg/archive/2000/468/cu3.htm [Accessed 20 November, 2017].

Al-Ahram-Weekly. 2000b. Derrida Perhaps. *Al-Ahram Weekly*, 17–23 February 2000, p. 469. http://weekly.ahram.org.eg/archive/2000/469/cu3.htm [accessed 20 November, 2017].

Al-Messirri, A. 2005. Derrida in Cairo: Deconstruction and Madness (Arabic ref.). *Philosophy Papers*, 12.

Alexander, C. 1979. *The Timeless Way of Building.* New York: Oxford University Press.

Alexander, C., Ishikawa, S. & Silverstein, M. 1977. *A Pattern Language: Towns, Buildings, Construction.* New York: Oxford University Press.

Alexander, H. G. (ed.) 1965. *The Leibniz-Clarke Correspondence: Together with Extracts from Newton's Principia and Optiks.* Manchester: Manchester University Press.

Allen, G. 2005. Intertextuality. *The Literary Encyclopedia* [Online]. Available: www.litencyc.com/php/stopics.php?rec=true&UID=1229 [Accessed 6 June 2010].

Alsayyad, N. 2006. Whose Cairo? *In:* Singerman, D. & Amar, P. (eds.) *Cairo Cosmopolitan: Politics, Culture, and Urban Space in the New Globalized Middle East.* Cairo, New York: The American University in Cairo Press.

Alvesson, M. & Skoldberg, K. 2000. *Reflexive Methodology: New Vistas for Qualitative Research.* London: Sage.

Amin, G. 2011. *Egypt in the Era of Hosni Mubarak 1981–2011.* Cairo: AUC Press.

Aristotle. 1984 [350 BC]. *The Complete Works of Aristotle: The Revised Oxford Translation.* Princeton, NJ: Princeton University Press (trans. J. Barnes).

Artaud, A. 1958. *The Theater and Its Double.* New York: Grove Press.

The Art History Archive. Russian Constructivism: Soviet Art. www.lilithgallery.com/arthistory/constructivism/ [Accessed 7 November 2013].

Badiou, A. 2005. *Metapolitics.* London: Verso (trans. J. Barker).

Barthes, R. 1976. *The Pleasure of the Text.* London: Cape (trans. R. Miller).

Bataille, G. 2001 [1962]. *Eroticism.* London: Penguin (trans. M. Dalwood).

Battesti, V. 2006. The Giza Zoo: Reappropriating Public Spaces, Reimagining Urban Beauty. *In:* Singerman, D. & Amar, P. (eds.) *Cairo Cosmopolitan: Politics, Culture, and Urban Space in the New Globalized Middle East.* Cairo, New York: The American University in Cairo Press.

Beeson, I. 1969. Cairo: A Millennial. *Saudi Aramco World.* September/October ed.

Bell, P. A. 2001. *Environmental Psychology.* Australia; United Kingdom: Thomson.

Bennington, G. 1992. Mosaic Fragment: If Derrida Were an Egyptian. . . . *In:* Wood, D. (ed.) *Derrida: A Critical Reader.* Oxford, UK; Cambridge, MA: Blackwell.

Bennington, G. 2001. Deconstruction Is Not What You Think. *In:* McQuillan, M. (ed.) *Deconstruction: A Reader.* New York: Routledge.

Bennington, G. 2003. Jacques Derrida. *In:* Culler, J. (ed.) *Deconstruction Critical Concepts in Literary and Cultural Studies.* London: Routledge.

Bennington, G. & Derrida, J. 1993. *Jacques Derrida*. Chicago: University of Chicago Press.

Berlin, I. 1969. *Four Essays on Liberty*. London, New York, etc.: Oxford University Press.

Berque, A. 1997. Postmodern Space and Japanese Tradition. *In:* Benko, G. & Strohmayer, U. (eds.) *Space and Social Theory: Interpreting Modernity and Postmodernity*. Oxford, England; Cambridge, MA: Blackwell Publishers, Special publications series (Institute of British Geographers); 33.

Berque, A. 2000. Overcoming Modernity, Yesterday and Today. *European Journal of East Asian Studies*, 1, 89–102.

Berque, A. 2005. Substantial Spaces, Existential Milieu: *l'espace ecoumenal* (French) *In:* Berthoz, A. & Recht, R. (eds.) *Les espaces de l'homme*. Paris: Odile Jacob.

Best, S. 2003. *A Beginner's Guide to Social Theory*. London; Thousand Oaks, CA: Sage.

Blackburn, S. 2008. *The Oxford Dictionary of Philosophy*. Oxford: Oxford University Press. http://www.oxfordreference.com.libproxy.ncl.ac.uk/ [accessed 23 November 2017].

Blackmore, S. 2005. *Consciousness: A Very Short Introduction*. Kindle Edition. Oxford: Oxford University Press.

Borg, I. 1978. Some Basic Concepts of Facet Theory. *In:* Lingoes, J.C. (ed.) *Geometric Representation of Relational Data: Readings in Multidimensional Scaling*. Ann Arbor: Mathesis Press.

Boullée, E.-L. 1968. *Essai sur l'Art*. Paris: Perouse de Montclos (Herman).

Broadbent, G. 1988. *Design in Architecture: Architecture and the Human Sciences*. London: Fulton.

Broadbent, G. 1990. *Emerging Concepts in Urban Space Design*. London; New York: Van Nostrand Reinhold (International).

Broadbent, G. 1991a. The Architecture of Deconstruction. *In:* Glusberg, J. (ed.) *Deconstruction: A Student Guide*. London: Academy Editions.

Broadbent, G. 1991b. The Philosophy of Deconstruction. *In:* Glusberg, J. (ed.) *Deconstruction: A Student Guide*. London: Academy Editions.

Broadbent, G. 1991c. Deconstruction in Action. *In:* Glusberg, J. (ed.) *Deconstruction: A Student Guide*. London: Academy Editions.

Broeck, P. V., Abdelwahab, M. A., Miciukiewicz, K. & Hillier, J. 2013. Analysing Space from a Strategic-Relational Institutionalist Perspective: The Case of the Cultural Park for Children in Cairo. *International Planning Studies IPS*, 18, 321–349.

Brown, J. 1985. An Introduction to the Uses of Facet Theory. *In:* Canter, D. V. (ed.) *Facet Theory: Approaches to Social Research*. New York: Springer.

Buckingham, W., King, P., Burnham, D., Weeks, M., Hill, C. & Marenbon, J. 2015. *The Philosophy Book*. London: Dorling Kindersley.

Canter, D. V., Ambrose, I., Brown, J., Comber, M., and Hirsch, A. 1980. Prison Design and Use. Unpublished Internal Report. Guildford: University of Surrey.

Canter, D. & Kenny, C. 1981. *The Multivariate Structure of Design Evaluation: A Cylindrex of Nurses' Conceptualizations* [Online]. Psychology Press. Available: www.informaworld.com/10.1207/s15327906mbr1602_6 [Accessed 16 February 2017].

Canter, D. V. 1977a. *The Psychology of Place*. London: Architectural Press.

Canter, D. V. 1977b. Place and Placelessness by E Relph: Book review. *Environment & Planning*. B, 4, 118–120.

Canter, D. V. 1982. Facet Approach to Applied Research. *Perceptual and Motor Skills*, 55, 143–154.

Canter, D. V. 1983a. The Purposive Evaluation of Places: A Facet Approach. *Environment and Behavior*, 15, 659–698.

Canter, D. V. 1983b. The Potential of Facet Theory for Applied Social Psychology. *Quality and Quantity: International Journal of Methodology*, 17, 35–67.

Canter, D. V. 1985. *Facet Theory: Approaches to Social Research*. New York: Springer.

Canter, D. V. 1997. The Facets of Place. *In:* Moore, G. T., Marans, R. W. & Environmental Design Research, A. (eds.) *Advances in Environment, Behavior, and Design v.4: Toward the Integration of Theory, Methods, Research, and Utilization*. New York: Plenum in cooperation with the Environmental Design Research Association.

Canter, D. V. & Rees, K. 1982. A Multivariate Model of Housing Satisfaction. *Applied Psychology*, 31, 185–207.

Casati, R. & Varzi, A. 200. Events. *In:* Zalta, E. N. (ed.) *Stanford Encyclopedia of Philosophy*. Fall 2008 ed. https://plato.stanford.edu/archives/fall2008/entries/events/ [Accessed 24 November 2017].

Casey, E. S. 1997a. *The Fate of Place: A Philosophical History*. Berkeley: University of California Press.

Casey, E. S. 1997b. Smooth Spaces and Rough-Edged Places: The Hidden History of Place. *The Review of Metaphysics*, 51(2), 267–296.

Castells, M. 1977. *The Urban Question: A Marxist Approach*. London: E. Arnold.

Collins, J., Mayblin, B. & Appignanesi, R. 2005. *Introducing Derrida*. Thriplow: Icon.

Cullen, G. 1961. *Townscape*. London: The Architectural Press.

Culler, J. 1982. *On Deconstruction: Theory and Criticism after Structuralism*. Ithaca, NY: Cornell University Press

Culler, J. 2003a. Deconstruction. *In:* Culler, J. (ed.) *Deconstruction Critical Concepts in Literary and Cultural Studies*. London: Routledge.

Culler, J. (ed.) 2003b. *Deconstruction Critical Concepts in Literary and Cultural Studies*. London: Routledge.

Deleuze, G. & Guattari, F. L. 1983. *On The Line*. New York, NY: Semiotext(e) (trans. J. Johnston).

Deleuze, G. & Guattari, F. L. 1987. *Thousand Plateaus: Capitalism and Schizophrenia*. Minneapolis: University of Minnesota Press, Ebooks Corporation (trans. B. Massumi).

Deleuze, G. & Guattari, F. L. 1994. *What Is Philosophy?* London: Verso.

Derrida, J. 1981. *Dissemination*. Chicago: University Press (trans. B. Johnson).

Derrida, J. 1982. Différance. *In: Margins of Philosophy*. Chicago: University of Chicago Press (trans. A. Bass).

Derrida, J. 1986. Point de Folie: maintenant L'Architecture. *In:* Culler, J. (ed.) *Deconstruction Critical Concepts in Literary and Cultural Studies*. London: Routledge.

Derrida, J. 1991a. A Letter to a Japanese Friend. *In:* Kamuf, P. (ed.) *A Derrida Reader: Between the Blinds*. New York: Columbia University Press.

Derrida, J. 1991b. Of Grammatology. *In:* Kamuf, P. (ed.) *A Derrida Reader: Between the Blinds*. New York: Columbia University Press.

Derrida, J. 1992. After.rds: Or, at Least, Less Than a Letter About a Letter Less. *In:* Royle, N. (ed.) *Afterwords*. Tampere, Finland: Outside Books.

Derrida, J. 1993. *Aporias*, Stanford, CA: Stanford University Press (trans. T. Dutoit).

Derrida, J. 1994. The Deconstruction of Actuality: An Interview with Jacques Derrida. *In:* Culler, J. (ed.) *Deconstruction Critical Concepts in Literary and Cultural Studies.* London: Routledge.

Derrida, J. 1995. *Khōra. On the Name.* Stanford, CA: Stanford University Press (trans. T. Dutoit).

Derrida, J. 1997a. Chora. *In:* Derrida, J., Eisenman, P., Kipnis, J. & Leeser, T. (eds.) *Chora L Works: Jacques Derrida and Peter Eisenman.* New York: Monacelli Press.

Derrida, J. 1997b. *Of Grammatology.* Baltimore: Johns Hopkins University Press (trans. G. Spivak).

Derrida, J. 2001 [1978]. *Writing and Difference.* London: Routledge (trans. A. Bass).

Derrida, J. 2004 [1979]. *Positions.* London: Continuum (trans. A. Bass).

Derrida, J. & De Man, P. 1989. *Memoires: For Paul de Man.* New York: Columbia University Press (trans. C. Lindsay).

Descartes, R. 1970. *Philosophical Letters.* Oxford: Clarendon Press (trans. A. Kenny).

Descartes, R. 1984 [1644]. *Principles of Philosophy.* Dordrecht, Holland; Boston, MA, Hingham, MA: Reidel, Distributed by Kluwer Boston (trans. V. Miller and R. Miller).

Descartes, R. 2006 [1637]. *A Discourse on the Method.* Oxford: Ebooks Corporation; Oxford: Oxford University Press (trans. I. Maclean).

Do, E. Y.-L. & Gross, M. D. 2001. Thinking with Diagrams in Architectural Design. *Artificial Intellegence Review,* 15, 135–149.

Donald, I. 1985. The Cylindrex of Place Evaluation. *In:* Canter, D. V. (ed.) *Facet Theory: Approaches to Social Research.* New York: Springer.

Dovey, K. 2005. *Fluid City: Transforming Melbourne's Urban Waterfront.* London: Routledge & Sydney: UNSW Press.

Dutoit, T. 1995. Translating the Name. *In:* Derrida, J. & Dutoit, T. (eds.) *On the Name.* Stanford, CA: Stanford University Press.

El-Messiri, N. 2004. A Changing Perception of Public Gardens. *In:* Bianca, S. & Jodidio, P. (eds.) *Cairo: Revitalising a Historic Metropolis.* Turin: Umberto Allemandi & C. for Aga Khan Trust for Culture.

Elsheshtawy, Y. 2004. The Middle East City: Moving Beyond the Narrative of Loss. *In:* Elsheshtawy, Y. (ed.) *Planning Middle Eastern Cities: An Urban Kaleidoscope in a Globalizing World.* London: Routledge.

Encarta-Online-Dictionary. http://uk.encarta.msn.com/ [Accessed 10 June 2009].

Encyclopedia Britannica. 1998. Meta Theory. https://www.britannica.com/topic/metatheory [Accessed 1 December 2017].

Ethelston, S. 2016. Facts and Figures on Cairo. *Middle East Report.* Middle East Research and Information Project, MERIP.

Forty, A. 2000. *Words and Buildings: A Vocabulary of Modern Architecture.* New York: Thames & Hudson.

Foucault, M. 1974. *The Archaeology of Knowledge.* London: Tavistock Publications (trans. A. M. Smith).

Foucault, M. 1981. *The History of Sexuality.* Harmondsworth: Penguin Books (trans. R. Hurley).

Fraser, I. & Henmi, R. 1994. *Envisioning Architecture: An Analysis of Drawing.* New York: Van Nostrand Reinhold.

Furth, M. 1967. Monadology. *The Philosophical Review,* 76, 169–200.

Galilei, G. 2008 [1615]. *The Essential Galileo.* Indianapolis: Hackett Publishing Company Inc., Ebooks Corporation (trans. M. A. Finocchiaro).

Gannon, S. & Bronwyn, D. 2012. Postmodern, Post-Structural and Critical Theories. *In:* Hesse-Biber, S.N. (ed.) *The Handbook of Feminist Research: Theory and Praxis.* Los Angeles, London, New Delhi, Singapore, Washington, DC: Sage.

Gee, J. P. 2004. *An Introduction to Discourse Analysis: Theory and Method.* London: Routledge.

Georgakopoulou, A. & Goutsos, D. 2004. *Discourse Analysis: An Introduction.* Edinburgh: Edinburgh University Press.

Glusberg, J. (ed.) 1991. *Deconstruction: A Student Guide.* London: Academy Editions.

Grant, E. 1981. *Much Ado About Nothing: Theories of Space and Vacuum from the Middle Ages to the Scientific Revolution.* Cambridge: Cambridge University Press.

Groat, L. 1981. Meaning in Architecture: New Directions and Sources. *Journal of Environmental Psychology,* 1, 73–85.

Groat, L. 1995a. *Giving Places Meaning.* London: Academic Press.

Groat, L. 1995b. Introduction: Place, Aesthetic Evaluation and Home. *In:* Groat, L. (ed.) *Giving Places Meaning: Readings in Environmental Psychology.* London: Academic Press.

Groat, L. & Després, C. 1991. The Significance of Architecture Theory for Environmental Design Research. *In:* Zube, E. H. & Moore, G. T. (eds.) *Advances in Environment, Behavior, and Design.* New York; London: Plenum.

Groat, L. N. 1984. Architecture and the Crisis of Modern Science by A. Pérez-Gómez: Book Review. *Journal of Environmental Psychology,* 4, 183–187.

Groat, L. N. & Wang, D. 2002. *Architectural Research Methods.* Hoboken, NJ: John Wiley & Sons, Inc.

Grosz, E. A. 1995. *Space, Time, and Perversion: Essays on the Politics of Bodies.* New York: Routledge.

Grosz, E. A. 2001. *Architecture from the Outside: Essays on Virtual and Real Space.* Cambridge, MA: [Great Britain], MIT Press.

Gustafson, P. 2001. Meanings of Place: Everyday Experience and Theoretical Conceptualizations. *Journal of Environmental Psychology – Academic Press,* 21, 5–16.

Guttman, L. 1965. A Faceted Definition of Intelligence. *Studies in Psychology,* 14, 166–181.

Hakim, B. S. 1999. View Points: Urban Form in Traditional Islamic Cultures: Further Studies Needed for Formulating Theory. *Cities,* 16, 51–55.

Hakim, B. S. & Rowe, P. G. 1983. The Representation of Values in Traditional and Contemporary Islamic Cities. *Journal of Architecture Education,* 36, 22–28.

Hamza, N. & Abdelwahab, M. 2017. Realizing Sensory Urban Environments: Decoding Synthetic Realities with Urban Performance Simulation. *In:* Yamu, C., Poplin, A., Devisch, O. & De Roo, G. (eds.) *The Virtual and the Real in Planning and Urban Design: Perspectives, Practices and Applications.* Abingdon, NY: Rouledge, Taylor and Francis Ltd.

Harvey, D. 1973. *Social Justice and the City.* Baltimore: Johns Hopkins University Press.

Hassan, F. 1997. The Cultural Garden in Sayyida Zeinab: An Expensive Toy – interview with Dr Abdelhalim Ibrahim Abdelhalim. Middle East Report.

Hattab, H. 2007. Concurrence or Divergence? Reconciling Descartes's Physics with His Metaphysics. *Journal of the History of Philosophy,* 45(1), 49–78. Project MUSE, doi:10.1353/hph.2007.0008.

Heidegger, M. 2002 [1927]. *On Time and Being.* London: University of Chicago Press (trans. J. Stambaugh).

Hejduk, R. 2007. Death Becomes Her: Transgression, Decay, and Eroticism in Bernard Tschumi's Early Writings and Projects. *The Journal of Architecture*, 12, 393–404.

Hill, L. 2007. *Jacques Derrida*. Cambridge: Cambridge University Press.

Hillier, B. 1996. *Space Is the Machine: A Configurational Theory of Architecture*. Cambridge: Cambridge University Press.

Hillier, B. & Hanson, J. 1988. *The Social Logic of Space*. Cambridge, UK; New York: Cambridge University Press.

Hillier, J. 2005. Straddling the Post-Structuralist Abyss: Between Transcendence and Immanence. *Planning Theory*, 4, 271–299.

Holland, R. 1999. Reflexivity. *Human Relations*, 52, 463–482.

Hollier, D. 1992. *Against Architecture: The Writings of Georges Bataille*. Cambridge, MA: MIT Press (trans. B. Wing).

Horrocks, C. & Jevtic, Z. 1997. *Foucault: For Beginners*. Cambridge: Icon Books.

Irigaray, L. 1985. *Speculum of the Other Woman*. Ithaca, NY: Cornell University Press (trans. G. Gill).

Ismaili-Studies, T. I. O. 2007. *Introduction to His Highness the Aga Khan* [Online]. The Institute of Ismaili-Studies. Available: www.iis.ac.uk/SiteAssets/pdf/aga-khan.pdf [Accessed 29 January 2009].

Johnson, P. & Wigley, M. 1988. *Deconstructivist Architecture*. New York: Museum of Modern Art.

Johnson, P.-A. 1994. *The Theory of Architecture: Concepts, Themes & Practices*. New York: Van Nostrand Reinhold.

Johnstone, B. 2008. *Discourse Analysis*. Oxford: Blackwell.

Kamel, B. A. & Abdelwahab, M. A. 2006. *Place-Making: The Ideal and the Real. ArchCairo 3rd Conference: Approching Architecture: Taming Urbanism in the Decades of Transformation*. Cairo: ArchCairo.

Kant, I. 2008 [1781]. *The Critique of Pure Reason*. Auckland: The Floating Press, Ebooks Corporation (trans. by J. M. D Meiklejohn).

Kellett, P. & Hérnandez-Garcia, J. 2013. Introduction: Researching the Contemporary City. *In:* Kellett, P. & Hérnandez-Garcia, J. (eds.) *The Contemporary City: International Research Perspectives*. Bogota: CEJA (Centro, Editorial Javeriano).

Kessel, T. 2007. Welcoming the Outside: A Reading of Hospitality and Event in Derrida and Deleuze. *Reconstruction: Studies in Contemporary Culture*, 7.

Kristeva, J. 1984 [1974]. *Revolution in Poetic Language*. New York: Columbia University Press (trans. M. Waller).

Kristeva, J. 1986 [1969]. Word, Dialogue and Novel. *In:* Moi, T. (ed.) *The Kristeva Reader*. Columbia: Columbia University Press, 34–65.

Kruft, H.-W. 1994. *A History of Architectural Theory: From Vitruvius to the Present*. London; New York; Zwemmer: Princeton Architectural Press.

Kuhn, T. S. 1970. *The Structure of Scientific Revolutions*. Chicago: University of Chicago Press.

Kymalainen & Lehtinen. 2010. Chora in Current Geographical Thought: Places of Co-Design and Remembering. *Geografiska Annaler: Series B, Human Geography*, 92, 251–261.

Lang, J. T. 1987. *Creating Architectural Theory: The Role of the Behavioral Sciences in Environmental Design*. New York: Van Nostrand Reinhold Co.

Launter, P. 2003. A New Survey of Neoplatonism by F. Romano: *Book Review. The Classical Review (New Series)*, 53, 83–85.

Law, J. 2003. *Making a Mess with Method*. Lancaster: Centre for Science Studies, Lancaster University.

Lawrence, R. J. 1982. A Psychological-Spatial Approach for Architcetural Design and Research. *Journal of Environmental Psychology*, 2, 37–51.

Lawrence, R. J. 1989. Structuralist Theories in Environmental-Behavior-Design Research: Applications for Analyses of People and the Built Environment. *In:* Zube, E. H. & Moore, G. T. (eds.) *Advances in Environment, Behavior, and Design*. New York; London: Plenum.

Lefebvre, H. 1991 [1974]. *The Production of Space*. Oxford, UK; Cambridge, MA: Blackwell (trans. D. Nicholson-Smith).

Levy, S. 2005. Guttman, Louis. *Encyclopedia of Social Measures*, 2, 175–188.

Locke, J. 1975 [1696]. *An Essay Concerning Human Understanding*. Oxford: Clarendon Press (trans. P. Nidditch).

Lucy, N. 2004. *A Derrida Dictionary*. Oxford: Blackwell.

Lynch, K. 1960. *The Image of the City*. Cambridge, MA: Technology Press.

Macarthur, J. 1993. Experiencing Absence: Eisenman and Derrida, Benjamin and Schwitters. *In:* Macarthur, J. (ed.) *Knowledge and/or/of Experience*. Brisbane, Queensland: Brisbane Institute of Modern Art.

Malisoff, W. 1940. What Is a Monad? *Philosophy of Science*, 7, 1–6.

Mallgrave, H. F. 2006. *Architectural Theory: An Anthology from Vitruvius to 1870*. Malden, MA: Blackwell Pub.

Markus, T. A. 1982a. *Order in Space and Society: Architectural Form and Its Context in the Scottish Enlightenment*. Edinburgh: Mainstream.

Markus, T. A. 1982b. Building for the Sad, the Bad and the Mad in Urban Scotland 1780–1830. *In:* Markus, T. A. (ed.) *Order in Space and Society: Architectural Form and Its Context in the Scottish Enlightenment*. Edinburgh: Mainstream.

Markus, T. A. 1986. Brief Encounter. *Building Design*, 12, 26–27.

Markus, T. A. 1987. Buildings as Classifying Devices. *Environment and Planning B: Planning and Design*, 14, 467–484.

Markus, T. A. 1988. Gardens of Delight and Deception. *Journal of New Blackfriars*, 69, 306–356.

Markus, T. A. 1993. *Buildings & Power: Freedom and Control in the Origin of Modern Building Types*. London; New York: Routledge.

Markus, T. A. & Cameron, D. 2002. *The Words Between the Spaces: Buildings and Language*. London: Routledge.

Martin, L. 1990. On the Intellectual Origins of Tschumi's Architectural Theory. *Assemblage*, 11, 22–35.

Matthews, P. H. 1997. *The Concise Oxford Dictionary of Linguistics*. Oxford: Oxford University Press.

McCormack, D. P. 2008. Thinking-Spaces for Research-Creation. *In:* Thain, A., Brunner, C. and Prevost, N., (eds.) How Is Research-Creation? *Inflexions: A Journal for Research Creation*, 1(1), 1–16. http://www.inflexions.org/

McDonough, J. K. 2014. Leibniz's Philosophy of Physics. *In:* Zalta, E. N. (ed.) *Stanford Encyclopedia of Philosophy*. Spring 2014 ed. https://plato.stanford.edu/archives/spr2014/entries/leibniz-physics/ [Accessed 24 November 2017].

McEwen, I. K. 2003. *Vitruvius: Writing the Body of Architecture*. Cambridge, MA: MIT Press.

McQuillan, M. 2001. *Deconstruction: A Reader*. New York: Routledge.

McQuillan, M. 2010. Aesthetic Allegory: Reading Hegel After Bernal. *In:* McQuillan, M. & Willis, I. (eds.) *The Origins of Deconstruction*. Basingstoke: Palgrave Macmillan.

Mendell, H. 1987. Topoi on Topos: The Development of Aristotle's Concept of Place. *Phronesis, Brill*, 32, 206–231.

Michele, C. 2009. Twenty Years After (Deconstructivism): An Interview with Bernard Tschumi. *Architectural Design*, 79, 24–29.

Miller, J. H. 2001. Derrida's Topographies. *In:* McQuillan, M. (ed.) *Deconstruction: A Reader.* New York: Routledge.

Mills, S. 2003. *Michel Foucault.* London: Routledge.

Moore, G. T. 1997. Toward Environment-Behavior Theories of the Middle Range: I. Their Structure and Relation to Normative Design Theories. *In:* Moore, G. T. & Marans, R. W. (eds.) *Advances in Environment, Behavior, and Design: Toward the Integration of Theory, Methods, Research, and Utilization.* New York; London: Plenum Published in Cooperation with the Environmental Design Research Association.

Morgan, E. 2006. *Derrida's Garden.* Vancouver: Projectile Publishing Society.

Mugerauer, R. 1992. Toward an Architectural Vocabulary: The Porch as a Between. *Dwelling, Seeing, and Designing: Toward a Phenomenological Ecology*, 103–128.

Mugerauer, R. 1994. *Interpretations on Behalf of Place: Environmental Displacements and Alternative Responses.* State University of New York Press.

Mullarkey, J. 2006. *Post-Continental Philosophy: An Outline.* London; New York: Continuum.

Newton, S. I. 1802. *Mathematical Principles of Natural Philosophy.* Printed by A. Strahan for T. Cadell & W. Davies. London: In the Strands. [trans. Thorp, R.]

Nilsson, F. 2004. Philosophy and the Development of Architectural Thinking. *In:* Cunnigham D. and Goodbun J. *Philosophy of Architecture. Architecture of Philosophy.* Bradford, UK: Centre for Cultural Analysis, Theory and History (CATH) Congress Proceedings, 9–11 July 2004.

Norris, C. 2004. Metaphysics. *In:* Reynolds, J. & Roffe, J. (eds.) *Understanding Derrida.* New York; London: Continuum.

Northrop, F. S. C. 1946. Leibniz's Theory of Space. *Journal of the History of Ideas,* 7, 422–446.

Nussbaum, M. C. 2001. *Upheavals of Thought: The Intelligence of Emotions.* Cambridge: Cambridge University Press.

Okasha, S. 2002. *Philosophy of Science: A Very Short Introduction.* Oxford: Oxford University Press.

Pearsall, J. 2001. *The Concise Oxford Dictionary.* Oxford: Oxford University Press.

Pérez-Gómez, A. 1994. Chóra: The Space of Architectural Representation. *In:* Pérez-Gómez, A. & Parcell, S. (eds.) *Chóra: Intervals in the Philosophy of Architecture.* Montreal, London: McGill University, History and Theory of Architecture Graduate Program; McGill-Queen's University Press.

Pérez-Gómez, A. & Parcell, S. (eds.) 1994. *Chóra: Intervals in the Philosophy of Architecture.* Montreal; London: McGill University, History and Theory of Architecture Graduate Program; McGill-Queen's University Press.

Plato 1892 [380 BC]. *The Republic.* Auckland: The Floating Press, Ebooks Corporation. (trans. B. Jowett).

Plato 1937 [360 BC]. *Plato's Cosmology: The Timaeus of Plato.* Whitefish, MT: Kessinger Publishing (trans. F. Cornford).

Potter, J. 1996. *Representing Reality: Discourse, Rhetoric and Social Construction.* London: Sage.

Pries, L. 2005. Configurations of Geographic and Societal Spaces: A Sociological Proposal Between 'methodological nationalism' and the 'spaces of flows'. *Global Networks,* 5, 167–190.

Rabbat, N. 2004. A Brief History of Green Spaces in Cairo. *In:* Bianca, S. & Jodidio, P. (eds.) *Cairo: Revitalising a Historic Metropolis.* Turin: Umberto Allemandi & C. for Aga Khan Trust for Culture.

Rakha, Y. 2001. Gyrating Intellects. *Al-Ahram Weekly*, 3 January 2001.

Rashed, D. 2005. To Salvage a City. *Al-Ahram Weekly*, 13–19 January 2005.

Raymond, A. 2007. *Cairo: City of History.* Cairo: The American University in Cairo Press.

Relph, E. 1976. *Place and Placelessness.* London: Pion Ltd.

Relph, E. 1978. The Psychology of Place by D. Canter: Book Review. *Environment and Planning A*, 10, 237–238.

Rickert, T. 2007. Toward the Chōra: Kristeva, Derrida, and Ulmer on Emplaced Invention. *Philosophy and Rhetoric*, 40, 251–273.

Rodenbeck, M. 2005. *Cairo: The City Victorious.* Cairo: AUC Press.

Rosenau, P. M. 1992. *Post-Modernism and the Social Sciences: Insights, Inroads, and Intrusions.* Princeton, NJ: Princeton University Press.

Rossi, A. 1981. *A Scientific Autobiography.* Cambridge, MA: MIT Press.

Routledge. 2000. *Concise Routledge Encyclopedia of Philosophy.* London: Routledge.

Royle, N. 2000. *Deconstructions: A User's Guide.* Basingstoke: Palgrave Macmillan.

Rozemond, M. 1995. Descartes's Case for Dualism. *Journal of the History of Philosophy*, 33, 29–63.

Rozemond, M. 1999. Descartes on Mind-body Interaction: What's the Problem? *Journal of the History of Philosophy*, 37, 435–467.

Russell, B. 1945. *A History of Western Philosophy and Its Connection with Political and Social Circumstances from the Earliest Times to the Present Day.* Stanford; New York: American Book.

Saegert, S. & Winkel, G. H. 1990. Environmental Psychology. *Annual Review of Psychology*, 41, 441–447.

Said, E. W. 2003. *Orientalism.* London: Penguin.

Salama, V. 2004. Rooftop Escape (Arabic ref.). *Al-Ahram Weekly*, 15 September.

Saleh, H. 1989. Abdelhalim I. Abdelhalim. *Cairo Today.*

Salheen, M. A.-K. 2001. *A Comprehensive Analysis of Pedestrian Environments: The Case of Cairo City Centre.* PhD, Heriot-Watt University, Edinburgh College of Art.

Sarup, M. 1993. *An Introductory Guide to Post-Structuralism and Postmodernism.* New York: Harvester Wheatsheaf.

Saussure, F. D., Bally, C., Sechehaye, A. & Reidlinger, A. 1955 [1983]. *Course in General Linguistics.* London: Duckworth (trans. R. Harris).

Schmaltz, T. M. 2009. Descartes on the Extension of Space and Time. *Analytica*, 13, 113–147.

Schneekloth, L. H. 1987. Advances in Practices in Environment, Behavior, and Design. *In:* Zube, E. H. E. & Moore, G. T. E. (eds.) *Advances in Environment, Behavior, and Design.* New York: Plenum Press.

Scott, J. & Marshall, G. 2005. *A Dictionary of Sociology.* Oxford; New York: Oxford University Press.

Scruton, R. 2001. *Kant: A Very Short Introduction.* Kindle Edition. Oxford: Oxford University Press.

Seamon, D. 1982. The Phenomenological Contribution to Environmental Psychology. *Journal of Environmental Psychology*, 2, 119–140.

Seamon, D. 2000. A Way of Seeing People and Place: Phenomenology in Environment-Behavior Research *In:* Wapner, S. (ed.) *Theoretical Perspectives in Environment-Behavior Research: Underlying Assumptions, Research Problems, and Methodologies.* New York: Springer.

Seidman, S. & Alexander, J. C. 2000. *The New Social Theory Reader.* London: Routledge.

Shye, S. (ed.) 1978. *Theory Construction and Data Analysis in the Behavioural Sciences.* London: Jossey-Bass.

Sime, J. D. 1985. Creating Places or Designing Spaces? *In:* Groat, L. (ed.) *Giving Places Meaning: Readings in Environmental Psychology.* London: Academic Press.

Singerman, D. (ed.) 2009. *Cairo Contested: Governance, Urban Space, and Global Modernity.* Cairo: AUC Press.

Singerman, D. & Amar, P. (eds.) 2006a. *Cairo Cosmopolitan: Politics, Culture, and Urban Space in the New Globalized Middle East.* Cairo, NY: The America University in Cairo Press.

Sixsmith, J. 1986. The Meaning of Home: An Exploratory Study of Environmental Experience. *Journal of Environmenatl Psychology, Academic Press,* 6, 281–298.

Sloterdijk, P. 2009. *Derrida, an Egyptian: On the Problem of the Jewish Pyramid.* Cambridge: Polity.

Slowik, E. 1999. Descartes, Spacetime, and Relational Motion. *Philosophy of Science,* 66, 117–139.

Smith, D. W. 2009. Phenomenology. *In:* Zalta, E. N. (ed.) *The Stanford Encyclopedia of Philosophy,* Summer. Available: http://plato.stanford.edu/archives/sum2009/entries/phenomenology/ [Accessed 20 October 2010].

Sokal, A. 1996. Transgressing the Boundaries: Towards a Transformative Hermeneutics of Quantum Gravity. *Social Text,* 46/47, 217–252.

Sollers, P. 1968 [1983]. *Writing and the Experience of Limits.* Columbia: Columbia University Press (trans. P. Barnard & D. Hayman).

Soltan, M. 1996. Mark Wigley: The Architecture of Deconstruction: Derrida's Haunt (Book Review). *Journal of Architecture Education,* 49, 266–268.

Solzhenitsyn, A. 1998. *Architecture and Deconstruction: A Brief Critique. Inland Architect: Looking for America, Part II: Decentering/Dislocation.* Chicago: Association of Collegiate Schools of Architecture (ACSA).

Stevenson, A. 2010 Oxford Dictionary of English, Oxford University Press http://www.oxfordreference.com/view/10.1093/acref/9780199571123.001.0001/acref-9780199571123 [Accessed 21 November 2017].

Sullivan, L. H. 1896. The Tall Building Artistically Considered. *Lippincott's Magazine,* 57, 403–409.

Thompson, J. 2008. *History of Egypt: From Earliest Times to the Present.* Cairo: AUC Press.

Tschumi, B. 1974. Fireworks. *In:* Tschumi, B. and Goldberg, R. (eds.) *A Space: A Thousand Words.* London: Royal College of Art Gallery.

Tschumi, B. 1975a. The Architectural Paradox. *In:* Tschumi(2001), B. (ed.) *Architecture and Disjunction.* 6th ed. Cambridge, MA: MIT Press – Reprint – first published as 'Questions of Space: The Pyramid and the Labyrinth (or the Architectural Paradox)' – Studio International September 1975, the Medical Tribune Group, London.

Tschumi, B. 1976. Architecture and Transgression. *In:* Tschumi(2001), B. (ed.) *Architecture and Disjunction.* 6th ed. Cambridge, MA: MIT Press – first published in Oppositions 1, Winter 1976, Cambridge: MIT Press.

Tschumi, B. 1977. The Pleasure of Architecture. *In:* Tschumi(2001), B. (ed.) *Architecture and Disjunction.* 6th ed. Cambridge, MA: MIT Press – Reprint – first published in Architectural Design, Academy Group Limited, London.

Tschumi, B. 1978. Architecture and Its Double. *In:* Tschumi(1990), B. (ed.) *Questions of Space: Lectures on Architecture*. London: AA publications – first published in Architectural Design, April.

Tschumi, B. 1980. Architecture and Limits I. *In:* Tschumi(2001), B. (ed.) *Architecture and Disjunction*. 6th ed. Cambridge, MA: MIT Press – first published in Artforum International, December 1980.

Tschumi, B. 1981a. Architecture and Limits II. *In:* Tschumi(2001), B. (ed.) *Architecture and Disjunction*. 6th ed. Cambridge, MA: MIT Press – first published in Artforum International, March 1981.

Tschumi, B. 1981b. Violence of Architecture. *In:* Tschumi(2001), B. (ed.) *Architecture and Disjunction*. 6th ed. Cambridge, MA: MIT Press – first published in Artforum International, September 1981.

Tschumi, B. 1983. Sequences. *In:* Tschumi(2001), B. (ed.) *Architecture and Disjunction*. 6th ed. Cambridge, MA: MIT Press – Reprint – published in The Princeton Journal: Thematic Studies in Architecture, Vol 1.

Tschumi, B. 1984. Madness and the Combinative. *In:* Tschumi(2001), B. (ed.) *Architecture and Disjunction*. 6th ed. Cambridge, MA: MIT Press – Reprint – first published in Precis, New York: Columbia University Press.

Tschumi, B. 1987a. Abstract Mediation and Strategy. *In:* Tschumi(2001), B. (ed.) *Architecture and Disjunction*. 6th ed. Cambridge, MA: MIT Press – first published as 'Cinégramme Folie: Le Parc de La Villette', New York: Princeton Architectural Press.

Tschumi, B. 1987b. Disjunctions. *In:* Tschumi(2001), B. (ed.) *Architecture and Disjunction*. 6th ed. Cambridge, MA: MIT Press – first published in Perspecta 23: The Yale architecture Journal, January 1987.

Tschumi, B. 1989. De-, Dis-, Ex-. *In:* Tschumi(2001), B. (ed.) *Architecture and Disjunction*. 6th ed. Cambridge, MA: MIT Press – first published in Barbara Kruger & Phil Mariani (eds.), 'Remaking History', Seattle: Bay Press.

Tschumi, B. 1994b. *Manhattan Transcripts*. London: Academy Editions.

Tschumi, B. 2001. *Architecture and Disjunction*. Cambridge, MA: MIT Press.

Tschumi, B. 2004. *Event-Cities 3: Concept vs. Context vs. Content*. Cambridge, MA: MIT Press.

Tschumi, B. 2010. *Event-Cities 4*. Cambridge, MA, London: MIT Press.

Ulmer, G. L. 1994. *Heuretics: Logic of Invention*. Baltimore, MD: Johns Hopkins University Press.

Venturi, R. 1988. *Complexity and Contradiction in Architecture*. London: Butterworth Architecture.

Verran, H. 2001. *Science and an African Logic*. Chicago; London: University of Chicago Press.

Vining, J. & Stevens, J. J. 1986. The Assessment of Landscape Quality: Major Methodological Considerations. *In:* Smardon, R. C., Palmer, J. F. & Felleman, J. P. (eds.) *Foundations for Visual Project Analysis*. Toronto: John Wiley & Sons.

Vitruvius, P. 1960. *Vitruvius: The Ten Books on Architecture*. New York: Dover Publications.

Walker, E. 2006. *Tschumi on Architecture: Conversations with Enrique Walker*. New York: Monacelli Press.

Weiss, A. 2005. The Transnationalization of Social Inequality: Conceptualizing Social Positions on a World Scale. *Current Sociology*, 53.

Wigley, M. 1995. *The Architecture of Deconstruction: Derrida's Haunt*. Cambridge, MA: MIT Press.

Wolfreys, J. 2007. *Derrida: A Guide for the Perplexed*. London; New York: Continuum.

Wong, J. F. 2003. *Reading Tschumi* [Online]. City University of Hong Kong. Available: http://personal.cityu.edu.hk/~bsjwong/tschumi.htm [Accessed 5 June 2010].

Wooffitt, R. 2005. *Conversation Analysis and Discourse Analysis: A Comparative and Critical Introduction*. London; Thousand Oaks, CA: Sage.

Yanow, D., Schwartz-Shea, P. & Ebooks Corporation. 2006. *Interpretation and Method: Empirical Research Methods and the Interpretive Turn*. Armonk: M.E. Sharpe Inc.

Young, J. 2002. *Heidegger's Later Philosophy*. Cambridge: Cambridge University Press.

Zepeda, J. R. 2009. *Descartes and His Critics on Space and Vacuum*. PhD, University of Notre Dame.

INDEX

Note: Page numbers in *italics* refer to figures.

Printed and bound by CPI Group (UK) Ltd, Croydon, CR0 4YY

24/10/2024

01778295-0005